The Climate Handprint

Gabriel Baunach

First published by Canbury Press 2025

This edition published 2025

Publisher: Canbury Press (www.canburypress.com)

14 Beresford Rd, London, KT2 6LR, United Kingdom

EU Authorised Representative: Easy Access System Europe

- Mustamäe tee 50, 10621 Tallinn, Estonia, gpsr.requests@easproject.com

Typeset in Centrale San Condensed (headings) and Athelas (body)

This translation of HOCH DIE HÄNDE, KLIMAWENDE first published
in Germany by Edition Michael Fischer GmbH in 2023 is published by
arrangement with Silke Bruenink Agency, Munich, Germany.

The information provided in this book is for informational purposes only and
does not constitute financial, legal, or professional advice. The author and
publisher are not licensed financial advisors, attorneys, or other professionals.
Readers should seek advice from qualified professionals regarding their unique
financial or legal situations. The author and publisher disclaim all liability for
any losses or damages incurred, directly or indirectly, as a result of the use or
reliance on the information provided. This book is sold with the understanding
that the author and publisher are not rendering professional services or advice of
any kind. Use of the information contained in this book is at your own risk.

This book was translated and revised with the help of AI tools like DeepL and
ChatGPT, and checked and overseen by the author and editor

ISBNs:

9781914487613 (paperback)

9781914487620 (epub)

The Climate Handprint

Gabriel Baunach

Canbury

In memory of Friedel Schütte (1933-2022), my grandfather, who gave me a feeling for the beauty of the Earth, its creatures and stories.

For my unborn children, his great-grandchildren, for whom I wish to preserve this beauty.

Contents

Introduction

One piece of bad, one piece of ambivalent and one piece of good news

Highway to hell

Sorry, I have to start this book with some bad news. United Nations Secretary-General Antonio Guterres put it like this:

> *'We are in the fight of our lives. And we are losing. [...] Our planet is fast approaching tipping points that will make climate chaos irreversible. We are on a highway to climate hell with our foot still on the accelerator...'* [1]

The fight that the United Nations (UN) chief diplomat is talking about is a race against time. If the tipping points of the global climate system are crossed due to global warming, critical ecological life-support systems will start to collapse – such as the Amazon rainforest, often called the 'green lungs of the Earth,' or the Gulf Stream in the Atlantic.[2] This can be compared to a garden pond tipping over in summer, but with catastrophic consequences on a *global* scale.[3]

The problem is that humanity is likely to trigger some of these dangerous climate tipping points with a global temperature rise of just over 1.5°C.[4] Among other things, the fish-rich warm-water corals would die off irretrievably. The permafrost in the far north of Alaska, Canada, Scandinavia, and Siberia – until now, permanently frozen – would thaw, releasing vast amounts of methane, a highly potent greenhouse gas. Meanwhile, the kilometre-thick ice sheets over Greenland and West Antarctica would melt inexorably, raising sea levels by several meters. That would mean 'Goodbye Fiji' in my lifetime, and my grandchildren would have to say 'Ciao Venice', 'Tschüss Hamburg', and 'Bye

London'. As the UN chief warns, the result would be a global 'mass exodus on a biblical scale'.[5]

The goal agreed upon by nearly all countries in Paris in 2015 is to limit the rise in global surface temperature to *well below* 2°C, ideally to 1.5°C. Given our current understanding of ecological and climate tipping points, it is clear that the Paris Climate Goals are not merely *goals*. Rather, they are *existential safety boundaries* for the survival of human civilization as we know it.

However, we are still miles away from meeting these existential safety boundaries.[6] Instead of heading towards 1.5°C, we are currently heading towards global warming of 2.7°C.[7] Despite all the climate promises, political targets and measures such as carbon pricing, coal phase outs, and an accelerating expansion of electric cars, heat pumps and renewable energies, far too little is still happening, far too slowly – even in wealthy countries such as the UK, Germany, and the US.[8]

The latest report by the Intergovernmental Panel on Climate Change (IPCC) states: 'There is a rapidly closing window of opportunity to secure a liveable and sustainable future for all.'[9] Or in the words of Antonio Guterres: 'We need all hands on deck for faster, bolder climate action. A window of opportunity remains open, but only a narrow shaft of light remains.'[10]

We have understood, but are on the wrong track

Now for an ambivalent message, which on the one hand gives optimism: By now, most people have understood the seriousness of the situation. 80 percent of people globally want stronger climate action by their government, according to a recent survey by the UN Development Programme and the University of Oxford.[11] Accordingly, a majority of Europeans support ambitious climate policies. And many are also prepared to make an active contribution themselves –

probably all those who pick up this book.

On the other hand, what dampens this fundamentally hopeful sense of optimism is the inadequate answer to the question of what each individual can do to meaningfully contribute to stabilizing our climate system. The vast majority of people answer: 'Reduce your personal carbon footprint.' But the problem is that the carbon footprint doesn't work – at least not well enough. Simply separating waste at home, turning down the heating, switching off the lights and cycling to the bakery on Sundays does not do justice to the scale, speed and structural causes of the climate crisis.

If we had another hundred years, small eco-friendly steps – like wooden toothbrushes, cloth bags, or organic vegetables – might be enough to make a difference. But time is running out. The United Nations Environment Programme (UNEP) has made it clear: **'Incremental change is no longer an option.'** Instead, what's needed is a **'wide-ranging, large-scale, rapid, and systemic transformation'** to keep our vital climate targets within reach.[12] This means that if we truly want to accelerate the climate transition on a societal level, we must focus on the big levers – politics, the legal system, the media, financial institutions, corporations, associations, environmental organizations, educational institutions, and more.

Unfortunately, the emphasis on individual carbon footprints distracts us from these critical systemic forces. It personalizes and depoliticizes what is fundamentally a collective and political crisis. As a result, we often get lost in the endless complexities of personal consumption, caught in frustrating debates over individual guilt, shame, and hypocrisy. Overwhelmed by our own footprints, we feel powerless, burdened by guilt, and stuck in place – or we turn against each other.

And that works perfectly for the oil companies. As long as we're busy blaming frequent flyers, SUV-driving uncles, or steak-loving friends, we're not pointing at the real culprits

– the fossil fuel industry, its shareholders and neoliberal networks, and the media companies and politicians they influence.

From footprint focus to handprint levers

The good news is that there is a solution. We just have to learn to recognize and use our collective power to shape social structures. 'Seeing the bigger picture' is what my grandmother used to call it. This book aims to do just that: That as many people as possible look up from their individual feet and recognize the levers with which they can bring about structural changes in their personal environment – and then become multipliers for climate-friendly behaviour with targeted handprint actions.

This book is for everyone who truly wants to make a difference but doesn't (yet) know how. For those who feel uncomfortable with the status quo but are waiting for the right push to get active. For those ready to embrace a fundamental shift in thinking. It is for all who want to be able to tell their children one day: 'Yes, I knew what was happening back then, and I did what I could to help solve humanity's greatest crisis. Thanks to the people who actively drove change in the 2020s, the world today, in 2050, is a better place.'

The book is called *The Climate Handprint* because we hold the power to shape the future in our hands (not so much under our feet): to create a more peaceful, healthier, fairer, better world. It's now, or never! All hands on deck! Let's go!

Chapter 1: My climate journey to the handprint

Give me a point where I can stand safely, a lever long enough, and I'll move the earth with one hand.
– Archimedes (285-212 BC)

I think the climate crisis is such a big problem that you need not one, but three 'aha moments' before you really take action. At least that was the case for me.

Al Gore wakes me up

I had my first big aha moment about the climate crisis at the age of 14 when we were shown the documentary *An Inconvenient Truth* in geography class. In this Oscar-winning documentary, Al Gore, the former Vice President of the USA, warns of the dangers of man-made climate change. In 2007, together with the Intergovernmental Panel on Climate Change (IPCC), he was awarded the Nobel Peace Prize for raising awareness of the impending consequences of our greenhouse gas emissions.

I remember exactly how I left the dark movie room of my school at the time, quite shocked. Instead of running to the table tennis table as usual during the break that followed, I wandered around the schoolyard, lost in thought. Up until that point, I had trusted that the world would get better and better and that I would live into a rosy future. I had believed that humanity would soon end the prevailing poverty, hunger, all wars and the overexploitation of nature. But my retrospectively naive foundation of trust began to crack as I listened to Al Gore's explanations about man-made global warming in the school playground. The inconvenient truth about the destabilization of the global climate startled me: I realized that we, humanity, face a huge problem that dwarfs all other problems, goals and initiatives – and that a rosy future without poverty, hunger or wars would become an illusion if we did not stop heating up our atmosphere.

Most people today feel the same way I did back then: 80 percent of British adults are very or fairly concerned about climate change and a large majority is also in favor of more ambitious climate policy measures.[13] As mentioned in the introduction to this book, this also applies globally: the need for stronger climate action is the absolute majority opinion.[14] The problem, however, is that we don't perceive it this way in the public discourse. Instead, we tend to greatly underestimate the willingness of our fellow human beings to support climate action.[15] And this has serious consequences: Many people behave in a less climate-friendly way or are more hesitant to support climate policy measures than they would actually be willing to.

Back to the school playground in 2007: I was gripped by the climate issue, but of course I had a lot of questions in my head after my first 'aha' moment. One of the reasons for this was that the topic was not yet given the space it deserved in the curriculum and our geography teacher didn't have time to explain climate change or solutions in more detail in the following lessons. So I had to go on a journey of discovery on my own. In the following years in my native Germany, I completed student internships at an energy and climate institute at Forschungszentrum Jülich and at the Potsdam Institute for Climate Impact Research, took an energy and climate course at the Deutsche SchülerAkademie during the summer vacation and read non-fiction books about climate change. And the more I learned, the firmer my resolve became: I want to help our species to survive. As I realized over time that renewable energies are the linchpin for solving the climate crisis, I decided to study mechanical engineering after graduating from high school and specialize in energy technology.

My student years involved stays abroad, more internships, some travel, football, parties, friends and more. The climate crisis faded into the background most of the time. Only now and then did the danger flash into my consciousness: For example, in 2014, when I saw with my own eyes the alarming glacier retreat on Ecuador's second highest mountain, the active volcano Cotopaxi, compared to my grandfather's travel photos and reports. Or in 2015, when the United Nations adopted the Paris Climate Agreement. In 2016, Leonardo DiCaprio enlightened me twice: first, with his appeal for climate action in his Oscar acceptance speech and, second, with his documentary Before the Flood. And in 2017, it was Al Gore's second climate documentary *An Inconvenient Sequel* that reminded me of the impending doom.

Flying less, eating less meat, cycling more, using green electricity and so on – I knew how to lead a climate-conscious life. However, this just lulled me into a false sense of security, which in retrospect can be described as naive, that humanity would solve the climate problem over time, with technological progress and without drastic changes.[16] A second disappointment was imminent.

The second awakening: my personal low point

My second big aha moment came shortly after the exceptionally hot and dry summer of 2018. While a then unknown teenager named Greta Thunberg went on school strike outside the Swedish parliament every Friday, the IPCC published its special report on the possibility of stopping global warming at a still relatively safe level of 1.5°C. (Whereby this 'relatively safe level' refers to our latitudes in the global North. For the global South, low-lying island states, indigenous communities in the Arctic, many ecosystems, etc., even 1.5°C of global warming will have catastrophic effects.)

Among other things, the report contains a graphic similar to the following figure. It shows how quickly and radically humanity would have to reduce global CO2 emissions in order to meet the 1.5°C target of the Paris Climate Agreement: global CO2 emissions would have to be halved by 2030 and reach net zero by 2050. But that's not all: after 2050, we would have to remove several billion tons of carbon from the air and put all this carbon back underground (for example by means of ecological restoration and afforestation).

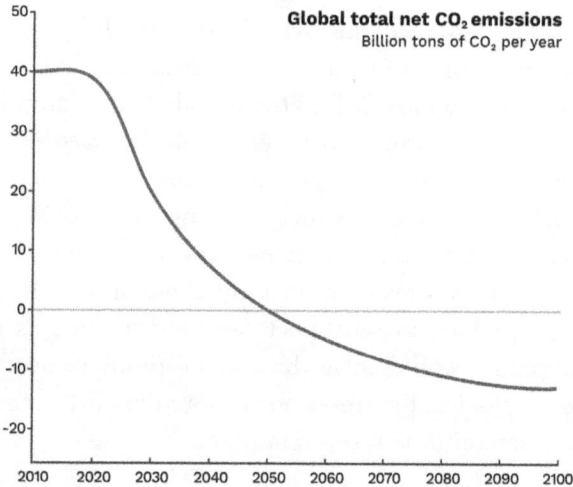

Global total net CO_2 emissions
Billion tons of CO_2 per year

Illustration 1: Necessary course of future CO2 emissions to limit global warming to 1.5°C (own illustration, based on the IPCC Special Report of 2018)

This graphic from the IPCC special report from 2018 burned itself into my memory. 'How on earth are we going to manage that?' I thought to myself and tried to behave even more climate-friendly. But it didn't help: fear, hopelessness and powerlessness spread through me as I joined, sometimes teary-eyed, thousands of children and young people in the streets to protest for their future under the banner of Fridays

For Future. A lost and dystopian future – that's what I thought at the time. Then the mixture of great climate anxiety, deep exhaustion after the tough mechanical engineering studies and my grandfather's fight for survival after his first stroke finally became too much for me: I was plagued by dizzy spells and I fell into an emotional hole that forced me to rest for several weeks.

I am not alone with my climate or eco anxiety. The majority of young people are feeling sadness, fear and worry in the face of the climate crisis – and the trend is rising.[17] There is also disillusionment among adults of all ages. In Britain, for instance, one third of people felt that any climate friendly changes that they made would have no effect.[18] This is no wonder: on the one hand, we are experiencing the first harbingers of the climate crisis, sometimes even first-hand: burning forests, dried-up rivers, overwhelming heat waves, deadly flood disasters and much more. On the other hand, politicians are still implementing far too few measures for a future worth living far too slowly, so that large parts of society simply continue to live as before. And in the midst of this depressing situation, we are just supposed to recycle, switch to bicycles, buses and trains, take shorter showers and switch off the lights whenever possible? If we are figuratively confronting the global conflagration known as the climate crisis with only toy water pistols, then it is not surprising that our dominant feelings are fear and powerlessness.

The third awakening: reunion with Al Gore and the discovery of the handprint

After unplugging for a few weeks in spring and early summer 2019, my vertigo attacks decreased significantly. Just in time for the start of my long-awaited internship at the UN Climate Change Secretariat in Bonn, I felt my energy returning. Luckily, my work at the United Nations – and in particular my participation in the 25th UN Climate Change Conference

COP25 – woke me up for a third time. Al Gore was there again, twelve years after he had given me my first aha moment in the movie room at my school. This time during the climate conference in Madrid, I was able to speak face to face with my childhood idol.

Through all my experiences and insights in the second half of 2019, I understood that the really influential levers for solving the climate crisis are not so much to be found in our homes at the garbage can, shower tap or light switch. My brief insight into the workings of global politics and economics made it clear to me what I had already suspected: that the really important levers lie first and foremost in politics – for example in international treaties and agreements, in the pricing of emissions, in laws, incentives, subsidies and bans – but also in the economy – namely in multinational corporations and large financial institutions and their investments, their research and in the development, production and advertising of goods and services. However, this realization does not mean that civil society – i.e. each and every one of us – is absolved of responsibility. Not at all. Because what I also realized was that the effective levers in politics and business can only be set in motion at the required speed if as many people as possible vehemently demand this and actively get involved.

So my third aha moment opened my eyes once again: no one is going to do it for us. Only we can save ourselves. But we're not going to do this with small acts and more climate-friendly (consumer) decisions such as wooden toothbrushes, cloth bags or organic vegetables. It is no longer enough to do without plastic bags or eat a little less meat. The IPCC puts it as follows in its latest major climate report: 'Individual behavioural change is insufficient for climate change mitigation unless embedded in structural and cultural change.'[19] In concrete terms, this means that we must use our work and free time, our skills and talents, our contacts and networks, our money and assets, our political voice and democratic rights to bring about this 'structural and cultural change' in time. With such levers, we can have a much greater influence on the economy and politics than with more climate-conscious consumer behaviour, exaggerated eco-morality or ascetic renunciation. It is time to take action in this way, i.e. to increase our handprint instead of just reducing our carbon footprint.

The Climate Handprint

In the early 2000s, 10-year-old Srija from Hyderabad in India took part in a project organized by the Indian Centre for Environment Education. In front of an audience, she said that she wanted to act positively and do more good instead of less bad for future sustainability. Thus the idea of the 'handprint' for sustainable action was born. Unlike the footprint, which usually indicates how much damage you do to the environment or how much bad you leave behind in the world, the handprint is intended to be a measure of positive and creative action – in other words, the good you do in the world. The handprint is therefore an optimistic and motivating alternative to the footprint.

The handprint concept first gained international recognition when it was presented at the fourth UNESCO conference on environmental education in Ahmedabad, India, in 2007. From then on, the idea spread to Europe accompanied by Srija's children's hand. The German non-profit organization Germanwatch played a major role in this, developing the handprint into a strategic concept for collectively changing structural conditions in one's own environment. Many of my explanations, thoughts and examples in this book are based on the handprint concept developed by Germanwatch (see information material in the appendix).[20]

Illustration 2: Srija's handprint became the symbol of the "handprint" for Germanwatch (photo credits: Germanwatch e.V., Benjamin Bertram)

I personally came across the handprint concept in 2020, when my climate education platform *Climaware* was already up and running. It immediately clicked, because my third awakening in 2019 led me to exactly the perspective that Srija already had at the age of 10: to work for more good and overall social change by increasing my own handprint and taking positive action, instead of just trying to do less bad in my everyday life by reducing my footprint.

I focused this positive and holistic perspective on the climate issue and whenever people asked me what they could do about the climate crisis, I would reply: 'Increase your climate handprint.'[21] This is because, on the one hand, increasing your own climate handprint can achieve far greater greenhouse gas savings and contribute much more to curbing global warming than reducing your personal carbon footprint. And secondly, the question 'What can I increase and where can I leave something good behind?' is a positive, inviting and motivating perspective on our individual options for action in the climate crisis. In contrast, emphasizing the carbon footprint leads to unpleasant feelings such as guilt or shame as well as negative framings such as reduction or renunciation.

For the sake of completeness, it should be mentioned that Harvard lecturer and life-cycle analysis expert Gregory Norris was working on developing the handprint concept for positive action at the same time as Germanwatch. In 2013, he published a paper titled *An Introduction to Handprints and Handprinting*, and in 2024, he gave a TEDx talk about his understanding of handprints at York Beach in the US.[22]

How the handprint effect works (exponential)

So far, the concept of the climate handprint is certainly still somewhat abstract. So let's imagine, for example, that Mary from Scotland makes a huge effort for a year to reduce as many CO_2 emissions as possible in her everyday life. She becomes a vegetarian, sells her car, stops flying on vacation, switches to a green electricity tariff, buys organic, regional and seasonal products whenever possible, buys her clothes from second-hand stores, separates her waste and makes sure she saves as much energy as possible at home. With her efforts, a few additional costs and a little extra time, she manages to reduce her carbon footprint by an impressive five tons. (For comparison: that's about eight flights from London to Mallorca and back or three years of driving a mid-range petrol car).[23] She thinks this is her maximum possible contribution to curbing climate change.

In reality, however, she could achieve tens, hundreds or even thousands of times more if she were to use her handprint levers in her job, in society or in politics. Because the impact of her personal climate action grows exponentially the more people she reaches and influences with her own actions, or how many other lives are changed by her actions. If she not only asked herself 'How can I live in a more climate-friendly way?', but also asked herself 'How can *as many of my fellow human beings as possible* live in a more climate-friendly way?', Mary would become a multiplier for climate-friendly behaviour. And that's exactly what the handprint is all about (see illustration below).

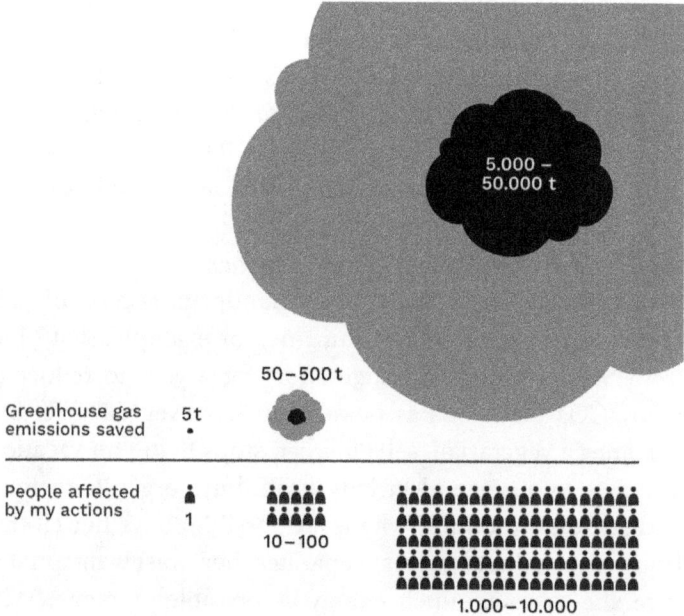

		5.000 – 50.000 t
	50–500 t	
Greenhouse gas emissions saved	5 t	
People affected by my actions	1	10–100
		1.000–10.000

Illustration 3: Schematic representation of the exponential effect of the climate handprint

What exactly do I mean by 'handprint levers'? These are all kinds of ways of influencing society to promote climate-friendly behaviour.[24] For example, by making it easier, cheaper, more socially attractive or required by law. In other words, handprint levers can be used to bring about a structural change in the framework conditions in which we live together, making it easier for even more people to act sustainably and reduce their personal footprints. Such handprint levers are available to us wherever we have a direct or indirect impact on a large number of people – in other words, in four dimensions of our lives:

- Firstly, in **a private context**, for example with family, friends or personal investments.
- Secondly, in the **workplace**, for example with colleagues and employees, with management or with products and services for customers.
- Thirdly, in **society**, for example through voluntary work or donations, through membership of an association, in the neighborhood, in church, school, training, university or via the media.
- And fourthly, in **politics**, for example through elections, petitions, demonstrations, discussions with politicians, party membership or legal action.

I wrote that Mary could reduce greenhouse gases by a factor of ten, a hundred or even a thousand times more than with her personal footprint reduction alone – as shown in the figure above. But how exactly can such an impressive reduction in emissions be achieved? The following handprint examples and rough calculations below illustrate this:

Let's assume that John is a student at a university whose canteen serves lunch to around 1,000 students 200 days a year. Of the five dishes on offer every day, only one is vegetarian. It usually costs more than the other four meat dishes and is also offered at the back of the food counter. John sees great potential for reducing greenhouse gases here and joins forces with a friend and two fellow students from the student council to bring about an environmentally and climate-friendly change with a petition in the canteen. And after a few weeks, the management of the canteen actually responded to their demands and suggestions: From now on, two vegetarian and one vegan dish will be offered daily, all of which will always be sold at a lower price than the two remaining meat dishes. In addition, the vegetarian and vegan dishes will move to the prominently located serving stations in the canteen and the meat dishes will be served further back. Only a short time later, 500 students are suddenly eating vegetarian or vegan lunches instead of the original 200. As a result, the canteen avoids a total of around 60 tons of greenhouse gases per year.[25] That's as much as John causes with his life in six years – and 12 times more than Mary has achieved with her footprint reduction. (However, Mary will have to make an effort again next year to maintain her change in behaviour. On the other hand, the structural change in the canteen happens once and then becomes the new permanent standard).

Second example: John's mother Claudia works in a medium-sized company that consumes around 500,000 kilowatt hours of electricity per year. Every day, she looks through her office window on the fifth floor at the large, unused roof areas of the company's production halls, above which the air begins to shimmer as soon as the sun shines. Inspired by her son's handprint campaign in the university canteen, she talks to colleagues, does her own research and finally takes heart and presents her own handprint idea to the company management: to equip the roofs of the halls with photovoltaic (PV) modules. With a little patience, persistence and the broad support of her colleagues, including the sustainability manager, Claudia actually managed to convince the company management to have the proposed solar systems installed after a few months. The self-produced solar power subsequently covers around half of the company's electricity consumption, meaning that Claudia's handprint campaign saves around 56 tons of CO_2 per year.[26] That is more than 11 times Mary's carbon footprint reduction.

The company's sustainability manager is now seized by ambition because he also wants to write sustainability success on his banner. He knows from internal company figures that around 100 of the company's employees, mainly from the sales department, travel a distance like the one from London to Edinburgh and back by plane around ten times a year on average. This generates around 3 tons of CO_2 per person per year, which means a total of 300 tons for the frequent flyer workforce. The sustainability manager actually manages to convince the company management at the next meeting to reduce this enormous amount of emissions by introducing rules for face-to-face customer meetings, video conferences and employee railcards. And in the following year, the measures to reduce flights actually had an impressive effect: every second flight was avoided thanks to better scheduling and travel planning, more video calls or rail travel. This results in a CO_2 saving of around 150 tons – a factor of 30 compared to Mary's carbon footprint reduction.

The really important levers for climate stabilization lie in the area of policies. To illustrate this, let's take the following as a final example: Ellie is involved with 50 people at a local climate activism group. With protests, a large-scale petition and discussions with Bristol City Council and experts, among other things, the alliance succeeds within a year in persuading Bristol's local politicians to switch the standard electricity supply of publicly owned buildings (e.g. administrative buildings, professional fire department, schools, daycare centers, museums and municipal apartments) from standard electricity to 100 percent green electricity. Let's assume, the city of Bristol's building management is responsible for an annual electricity consumption of around 50 million kilowatt hours. This results in greenhouse gas emissions savings of around 11,250 metric tons per year if this electricity requirement is covered by carbon-neutral and renewable energies.[27] Ellie's handprint is therefore on the far right of the diagram, with the largest CO_2 cloud. It would be even larger if the local energy supplier were to switch the standard electricity tariff for *all* buildings in Bristol to 100 percent green electricity as a result of the group's climate commitment.

Between 2010 and 2012, the municipal utilities of St. Gallen in Switzerland proved that this is not a completely far-fetched idea from the realm of utopian dreams: Despite intensive marketing efforts, by 2010 they had only managed to convince a small proportion of households in the Swiss canton of St. Gallen to switch to a green electricity tariff. But then they reversed the choice architecture: Instead of having to switch to a green electricity tariff, green electricity was henceforth the standard and anyone who didn't want it had to actively switch back to a grey electricity tariff. This increased the proportion of green electricity customers from the original 10 percent to around 90 percent within just two years.[28] However, less drastic interventions in the decision-making architecture when choosing an electricity tariff also deliver impressive results: A 2015 study in Germany found that if online electricity portals simply tick the green electricity tariff box for their potential customers from the outset, the number of green electricity contracts concluded can increase tenfold.[29]

As can be seen from the examples above, pretty much everyone (depending on their circumstances) has many opportunities to make a collective contribution to the climate transition for society as a whole. But unlike footprint reduction, this hardly ever works alone. The handprint is almost always about working with people to influence other peoples' behaviour.

Pitfalls of the climate handprint – and how to mitigate them

Of course, the examples above are based on a large number of assumptions and estimates. In reality, however, it is extremely difficult (if not impossible) to calculate a person's exact climate handprint. The interdependencies and decision-making chains are too complex in most cases, which is why the final result cannot be precisely attributed to and credited to individual people. John's action group may have received support from the kitchen staff, Claudia's handprint may have involved her boss and the installers of the solar panels, the sustainability manager may have had support from the company's HR department in addition to the management floor and, in Ellie's case, the city council and the local energy supplier were also involved in the electricity tariff change. Ultimately, such changes are complex collective transformation processes. It is therefore only possible to estimate an approximate order of magnitude of the emissions reduction achieved for each individual involved.

When calculating the respective climate handprint in the examples, however, I am less concerned with the exact numbers than with the idea. And that is: Our handprint levers are many times more influential and effective than our individual footprint levers. And the more we make it easy for people to lead a climate-friendly life, the greater the potential CO_2 savings, or the higher the level of impact: from the private sphere to the workplace and parts of society to politics. (In startup jargon, the handprint scales exponentially with the level of impact).

If you have been paying close attention, you may have noticed that the greenhouse gas savings in the handprint examples above can be counted twice. John's handprint due to the change in the university canteen's menu is made up of the many individual carbon footprint reductions of his 300 or so fellow students, who are suddenly opting for more climate-friendly food. His mother Claudia only achieved a climate handprint of 56 tons because the carbon footprint of her colleagues in the office is automatically and imperceptibly reduced by solar power. And the sustainability manager's climate handprint of 150 tons of avoided greenhouse gases consists of the individual CO_2 savings of his 100 former colleagues who used to fly a lot.

We must always keep the following principle in mind: **Increasing my handprint means reducing someone else's footprint.** The key difference, however, is that the decisions leading to their footprint reductions are either taken off their shoulders or made easier for them. In other words, you don't have to painstakingly convince people to change their behaviour. Instead, you can remove climate-damaging obstacles or gently nudge them in the right direction – through pricing, choice architecture, and structural changes – so that they naturally adopt more climate-friendly behaviours on their own. As long as John isn't debating with a fellow student at the vegan food counter in the university canteen about whether the kilogram of greenhouse gases avoided in her lunch is due to *his* handprint expansion or her footprint reduction, there's no real issue with counting both the climate handprint and carbon footprint in parallel.

The climate handprint only becomes a problem when it serves as an excuse for not reducing one's own carbon footprint. For example, when Claudia and John decide to buy a new combustion-powered SUV for the family or plan a weekend trip to Barcelona by plane because they think it's okay in view of their exemplary handprint actions, they fall into psychological traps: They offset bad deeds with good ones, and part of their original environmental success fizzles out in the SUV exhaust or in the airplane turbine. However, this probably happens less often: Scientific studies indicate that socially committed climate activists with large climate handprints generally also behave in an increasingly climate-friendly manner in their private lives.[30]

Before we can address the question of how exactly each and every one of us can increase our climate handprint, we first need to dive deep into the mess that the excessive focus on our individual carbon footprints has created. We start at the very beginning: with the big oil companies and the dubious history of the spread of the footprint concept.

Chapter 2: With the carbon footprint from oil advertising into The Matrix

In the 1999 film *The Matrix*, the protagonist named Neo realises that what he has believed to be reality for all the years of his life is in fact an illusory world only for his mind. In the film, this illusory world is created by the 'Matrix', a kind of computer programme created by intelligent machines to keep people's minds trapped in an illusion. While people seem to lead a 'normal' life in the illusory world of the Matrix, in the real world their generated body energy is tapped in huge fields with cables and hoses as a source of energy for the machines. Only a fraction of humanity resists the rule and oppression of the intelligent machines. Neo can only recognize the real reality after he decides to swallow a red pill instead of a blue pill, offered to him within the Matrix illusory world. The red pill stands for recognizing the truth, the blue one for continuing to live in the illusion.

Just like in *The Matrix*, the vast majority of people who want to do something about the climate crisis are living in an illusion. This is because the answer to the question 'What can I do about the climate crisis?' is almost always 'Reduce my own carbon footprint' – and almost never 'Increase my climate handprint'. We are trapped in a kind of 'footprint matrix'. Never before have there been so many guides, apps and lists of recommendations on the subject. We all know the individual eco-tips by now: 'cycle more', 'eat less meat', 'use green electricity' and 'switch to public transportation'. On the other hand, there is very little information on how individuals can put pressure on politicians and corporations, promote structural change and increase their positive handprint. The result is our current green transition in slow motion.

The red pill I offer in this book stands for the following realisation: We will not stop global warming close to 1.5°C – and thus prevent the worst consequences of the climate crisis – if the people who want to be part of the solution continue to pay attention only to their individual carbon footprints. To believe that we can effectively curb the heating of our atmosphere by making more environmentally conscious consumer choices is an illusion. This is because the carbon footprint does not work – at least not well enough: it places an emotional burden on us, but it does not match up to the scale, speed and structural causes of the climate crisis.

How did it come about that almost all more or less eco-conscious people only think about their private consumption and energy consumption? How did the term 'carbon footprint' come about in the first place? Or to put it more conspiratorially: who wrote the source code for the footprint matrix – and why?

On the trail of the carbon footprint myth

My search for the origins of the carbon footprint began in late summer 2020, when Prof. Dr. Stefan Rahmstorf – head of department at the Potsdam Institute for Climate Impact Research and one of Germany's best-known climate scientists – astounded me with the following sentence during the recording of our podcast interview: '[What] has also been heavily pushed by oil companies is this idea of individual responsibility: meaning, the consumer is to blame.'[31] I found the idea as perfidious as it was ingenious: that oil companies were deliberately shifting their responsibility for the climate crisis onto the shoulders of individuals – our shoulders. In simple terms: we are to blame – not the oil bosses, fossil fuel lobby groups or the politicians they influence.

I looked into Stefan Rahmstorf's statement immediately after the interview. And sure enough, I found an article by journalist Mark Kaufman entitled *The Carbon Footprint Sham*.[32] It describes how one of the world's largest oil companies spread the term 'carbon footprint' to the public with a gigantic advertising campaign and its own carbon footprint calculator. The more research I did, the more I thought Stefan Rahmstorf was right.

The footprint of a computer on a desk in Canada

But let's start at the very beginning: at the origin of the footprint. More precisely, in the office of the now retired professor William Rees at the University of British Columbia in Canada. When the university put a new, space-saving computer on his desk in 1992, William Rees was working on a scientific paper on environmental issues. His doctoral student Mathis Wackernagel noticed the new PC and asked how he liked it. William Rees replied that he particularly liked the 'smaller footprint' of the smaller device on his desk. Then the idea struck him like lightning: the smaller desk footprint of his computer could also be transferred to the resource consumption of human activities. Environmental pollution could be represented as a footprint, an ecological footprint.[33] Within a few minutes, William Rees incorporated the neologism into his scientific paper and became the forefather of the concept.[34] (And his PhD student Mathis Wackernagel became the president of the Global Footprint Network, an international think tank for sustainability).[35]

When the problem of man-made global warming gained attention in the 1990s, ways were sought to communicate the very complex issue as simply as possible. Applying the concept of the ecological footprint to the climate problem was an obvious step. This was first done publicly in 2000 in a newspaper interview. However, it was not by William Rees or another person from the world of science, but by the chief environmental officer of a Texan electricity company. He said at the time that it was important at this point in world history that we reduce our carbon footprint as well as our ecological footprint.[36] In doing so, he unwittingly planted the seed for a large-scale propaganda campaign that made the term known everywhere.

The invention of climate greenwashing

At around the same time, the global oil industry was threatened by a lack of communication strategy. Even then the oil companies knew exactly what catastrophic effects the emissions would have and that climate change was man-made.[37] But they had publicly denied or doubted in articles, advertisements and press statements that mankind was responsible for the measured global warming or that global warming had negative consequences.[38] After the publication of the first two reports of the IPCC in the 1990s, that communication strategy of public denial and doubt was no longer credible. The scientific burden of proof for man-made global warming had become too great. So a new approach was needed to deal with the greenhouse gas problem in public, to 'capture' it communicatively. (However, the oil companies' campaigns of denial and doubt continued covertly via pseudo-scientific institutes and fossil fuel lobbying associations.[39])

Against this backdrop, BP (British Petroleum) – the world's second largest non-state-owned oil company – pioneered a new approach, with an open, almost humble and friendly campaign. In the early 2000s, BP commissioned the PR and advertising company Ogilvy & Mather to run a large marketing campaign that would present BP as an environmentally conscious company that wanted to play a part in solving the climate crisis. BP was renamed bp (with lower case letters), which would henceforth stand for 'beyond petroleum' and a new yellow-green sunflower logo replaced the martial shield logo. And it worked: between 2000 and 2007, brand awareness of bp rose from an initial 4 percent to 67 percent. A customer survey in 2007 also revealed that bp was perceived by the public as the most environmentally friendly oil company by a wide margin. In the same year, the beyond-petroleum campaign won the Gold Award from the US Marketing Association.[40]

At the same time, however, one of the central architects of the bp marketing campaign at Ogilvy & Mather, John Kenney, began to have serious doubts about the authenticity of the entire campaign. In the summer of 2006, he wrote an article for the *New York Times* entitled *Beyond Propaganda*.[41] In it, John Kenney described the euphoria he experienced at the beginning of the bp campaign. How exciting it seemed at the time to be able to initiate a real change towards clean energy sources in one of the world's largest oil companies. How motivating the street interviews with unprepared US citizens were, because in the early 2000s most people already wanted oil companies to take a leading role in environmental and climate protection.

Unfortunately, their will was not heard. On the contrary: in the ten years following the launch of the campaign, bp sold or terminated considerable parts of its already tentative commitment to renewable energies.[42] In addition, bp's green – *greenwashed* – image was tarnished by high-profile disasters: a fatal explosion at a Texas refinery in 2005, an oil spill in Alaska in 2006 and the Deepwater Horizon oil spill in the Gulf of Mexico in 2010.[43] Today we know: The slogan 'beyond petroleum' was nothing but hot air. Shortly before the coronavirus pandemic, bp was still producing roughly the same amount of oil and gas as in 2005. In 2022 the fossil fuel business brought the company the highest profits in its history and in 2023 bp even announced that it was lowering its climate targets for 2030.[44] I am therefore now deliberately switching back to the capital letters BP.

Here's the thing: it was this large-scale propaganda campaign, worth between 200 and 370 million US dollars depending on how you count it, that helped to popularise the term 'carbon footprint'.[45] Alongside the green livery with the renaming to 'beyond petroleum', and the numerous advertising videos, BP published its masterpiece at the time: one of the world's first personal carbon footprint calculators.[46] With full-page print ads in the most widely read newspapers in the USA (such as the *New York Times*) and TV commercials, BP brought the carbon footprint into the public eye and directed people to the calculator on the BP website.[47]

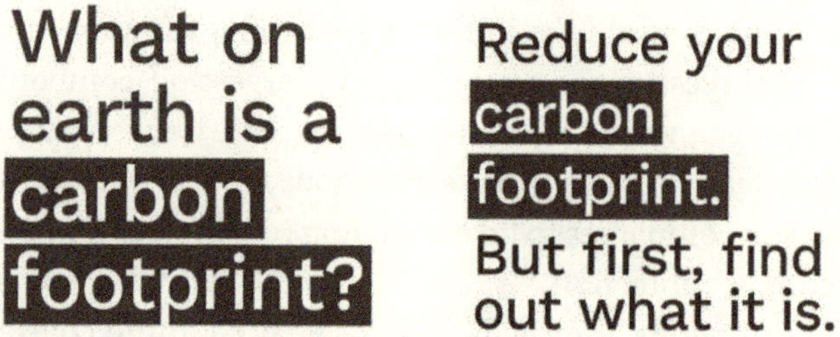

What on earth is a carbon footprint?

Reduce your carbon footprint. But first, find out what it is.

Illustration 5: BP advertising in the period 2004-2006

A person who calculates their annual carbon footprint using the BP tool may not be aware that the oil companies ExxonMobil, Shell and BP have been responsible for the most industrial greenhouse gas emissions of any non-state-owned company in the world since 1988.[48] In recent times, BP alone has been responsible for around 376 million tons of greenhouse gas emissions per year.[49] For comparison: If you equate an average individual carbon footprint of 10 tons with the actual area of a human footprint (approx. 250 cm^2), then the size of BP's footprint is equivalent to the area of around 130 football pitches or 18 Egyptian pyramids of Cheops side by side.[50]

Apparently, people loved the information about supposedly effective everyday actions against the climate crisis from an oil company: In 2004 alone – at a time when the Yahoo search engine still had the most traffic on the internet, ahead of Google – BP's website attracted a staggering 1.13 million visitors. And hundreds of thousands clicked through the brand new carbon footprint calculator in the first year of publication alone, as BP later proudly announced. For an oil company and the low level of internet usage at the time, these are astonishing figures.[51]

Today, Google's language analysis tools show us how effective the BP campaign was on the spread of the term carbon footprint: use of the term increased exponentially from 2004 in both English language and literature.[52] As a result, the perspective of individual consumer responsibility for climate protection received 'its biggest boost [...] from a huge BP marketing campaign about the carbon footprint'. This is how none other than Dr. Mathis Wackernagel, one of the two founders of the footprint concept, put it in an interview.[53]

Now, of course, one might ask why one of the world's largest oil companies wants to inform the public about the amount and climate damage caused by CO_2 emissions. Wouldn't silence and a cautious, defensive communication strategy have been wiser for BP? So what was BP's motivation behind 'beyond petroleum' and the massive promotion of the carbon footprint? I see two options:

Option 1: It was a well-meaning attempt to overcome the world's dependence on oil and gas, to forego short-term profits and to focus on long-term profits from renewable energies. If so, from today's perspective, the BP campaign would have been an authentic but failed spark of idealism.

Option 2: It was designed to distract millions of people worldwide from the destructive machinations of the company and oil industry. Then it would have been one of the cleverest, but also most perfidious strategies in marketing history.

We will probably never know what was really going on in BP management levels at the beginning of the 2000s. However, the overall picture that emerged from my research and the sources that are publicly available today points in the direction of option 2. And there is further evidence for this.

Distraction through blame shifting

One indicator is that marketing experts have long known that the art of subtly concealing and shifting blame can increase sales of products and services.[54] But that's not all: so-called 'guilt framing' can also be used to shift public attention from actors in harmful economic sectors – for example tobacco companies or fast-food chains – onto individuals. From the perspective of these companies, this distraction mechanism through subtle blame shifting is ingenious: because if individuals are mainly concerned with their consumption decisions, they collectively put less pressure on politicians for regulatory measures and laws. And as long as politicians do not take the producers of harmful products to task and regulate them, their profits continue to rise.

A macabre example is the guilt framing of the most powerful gun lobby in the USA, the National Rifle Association (NRA). Among other things, it coined the slogan 'Guns don't kill people, people kill people' – to distract from the fact that the more firearms there are in a country, the more gun deaths there are.[55] As a result, there is more discussion in the US about individuals' mental health problems and arming teachers than about stricter gun laws. And gun manufacturers continue to produce and sell guns.

But this distraction method also has a long tradition in environmental protection: in 1971, the Keep America Beautiful initiative in the USA launched one of the most successful advertising campaigns of all time. Among other things, the following advertising sequence was shown on US television: A seemingly indigenous man – in reality an Italian-born actor nicknamed 'Iron Eyes Cody' – paddles down a river in a canoe, which becomes increasingly polluted with garbage along the journey. Deeply affected and hurt by this pollution, a tear rolls down the cheek of 'Iron Eyes Cody'.[56] The image of the 'Crying Indian' went around the world. Behind the advertisement, however, were America's largest beverage and packaging manufacturers, under the aegis of the group Keep America Beautiful.[57] Their slogan 'People start pollution, people can stop it' is reminiscent of the NRA's slogan for a reason. It contains the hidden message: it's not Coca-Cola that is to blame for environmental pollution, it's *you*.

We see distraction through subtle and indirect shifting of blame from large corporations to individuals worked well even before the bp campaign. It's very hard to imagine BP's marketing and communications experts were not aware of this. Due to the very similar methodology, this leads to the conclusion that the individualization of the global, structural climate crisis by means of the call to 'reduce your carbon footprint' was also a targeted guilt framing like the NRA and 'Keep America Beautiful'.

A second point supports this conclusion: after the successful bp campaign, the other major oil companies – such as Shell, ExxonMobil and Chevron – also jumped on board BP's new communication strategy, but without fundamentally changing their business model from fossil to renewable. And this has still not changed to this day. Distraction and blame-shifting is still the dominant strategy of 'Big Oil', the major oil and gas companies. For example, a study in 2022 using a machine learning analysis of over 180 communication documents from ExxonMobil found that the company's PR department deliberately downplayed climate risks and shifted the blame and responsibility onto individual consumers through an excessive emphasis on consumption instead of production.[58] The oil giant Shell demonstrably used distracting and individualizing rhetoric in its advertising and public communications.[59] And BP has still been advertising its carbon footprint for around 20 years – including on the company's website and social media channels.[60] So the climate greenwashing continues apace. (In 2021, only 12 percent of Big Oil's – that's BP, Chevron, ExxonMobil, Shell and TotalEnergies – capital expenditure went into low-carbon activities and developments, while these five oil companies together spent around half a billion US dollars on 'green' public relations.)[61]

The fact that Big Oil companies are still cleverly distracting, deceiving and shifting the blame and responsibility onto individuals is due to the fact that they do not want to move 'beyond petroleum' at all, despite climate protection pledges and green promises for the future.[62] Firstly, this is demonstrated by their fossil fuel investment plans for the coming years, which alone would cause so many CO_2 emissions that global warming could no longer be halted at 1.5°C.[63] Secondly, the non-profit organization Oil Change International 2024 detailed in its *Big Oil Reality Check* report that not a single major oil company has reduced oil or gas production, set a phase-out date for fossil fuels that is somewhat aligned with the United Nations' Paris Climate Agreement or at least stopped deceptive advertising or climate-damaging lobbying.[64]

Fortunately, the climate movement has become more vigilant since 'beyond petroleum'. Communications from oil companies that overly allude to the responsibility of individuals are increasingly being exposed for what they mostly are: greenwashing, distraction and blame or responsibility shifting. Despite this increasing vigilance on the part of some committed individuals, the focus on the individual carbon footprint still dominates the majority of society. Instead of focusing on politicians and the big players in the climate crisis and trying to get them to change the structural framework conditions and rules of the game for the fossil fuel industry (for example through a higher and socially redistributed carbon tax), we still collectively point the finger at our aunt, neighbor or colleague and blame them for their car, their meat consumption, their last vacation trip or their plastic consumption. Studies show that precisely this behaviour – or the feeling of individualized responsibility – is one of the main reasons why people reject more ambitious climate policy measures (such as socially redistributed carbon taxes).[65] And this, of course, plays into the hands of all corporations that make money in any way from burning fossil fuels.

Caught in the footprint matrix: Blue pill or red pill?

'All I'm offering you is the truth, nothing more.' With these words, Neo's mentor gives him a choice in the movie *The Matrix*: blue or red pill? After a brief hesitation, Neo takes the red pill and then learns how the Matrix works ... Ready for the red pill to learn about the *footprint matrix*?

Climate action through carbon footprint reduction	Not doing enough	(Supposedly) doing enough
High awareness	2. Inwardly Torn	3. Temporarily Frustrated
Low awareness	0. Still Unwilling	1. Haphazard Doers

Figure 6: The footprint matrix

The footprint matrix can be presented in the form of a table with four fields along two axes. The vertical axis or the respective row of the table qualitatively indicates a person's *awareness of individual climate action through carbon footprint reduction*. Those with a high level of awareness already have above-average knowledge about the climate crisis and the opportunities to reduce their own carbon footprint. In contrast, those with a low level of awareness have relatively little knowledge about the climate crisis and do not understand exactly how they can save as much CO_2 as possible in their everyday lives. The horizontal axis or the respective column in the table describes the *perceived extent* of a person's action. Those who are still inactive – or feel that they are doing too little to reduce their per capita emissions – end up in the left-hand column: *Not doing enough*. On the other hand, those who are doing a lot – or just think they are already doing a lot – are in the right-hand column: *(Supposedly) doing enough*. We can now categorize most people who would be willing or are already trying to reduce their carbon footprint into one of the four fields of this matrix. Each field thus represents a 'footprint stereotype'.

Why I exclude the Still Unwilling

The table at the bottom left shows the 'Still Unwilling'. This type is an exception, as they have not (yet) grasped the extent, danger and urgency of the climate crisis, despite the ever-increasing heatwaves, forest fires, floods, droughts and so on. Although the danger of the climate crisis is now so obvious, this group doubts the findings of climate science, considers the whole 'climate hype' to be completely exaggerated, is therefore opposed to political climate protection measures and does not want any (climate-friendly) changes in their own lives. Some of them may still be open and receptive to climate protection arguments, but do not see themselves as responsible for taking action. For example, because they mistakenly focus solely on innovation and purely technical solutions or only see the cause of the problem in China or global 'overpopulation'. These few may still be convinced in the future that the climate crisis poses a real threat to them and their loved ones, that climate protection will ultimately benefit them too and that they also bear some of the responsibility. But the majority of those who are still unwilling are a tough nut to crack and many climate scientists and journalists, among others, have been trying to educate them for decades. They slow down, block and trivialize climate protection efforts whenever and wherever the topic comes up.

I am deliberately excluding the Still Unwilling from this book. The reason for this is that they are not unconvinced because there is a lack of information, facts or accuracy in the scientific findings. A large number of studies have shown that it does not help to repeatedly confront those who are unconvinced or unwilling to change with even more scientific facts.[66] The knowledge-deficit model that is still frequently practiced – including by many scientists, media people and environmental campaigns – is failing.[67] And those who are still unwilling often do not want any climate-relevant information at all and even actively avoid it.[68]

No, much more than the amount, accuracy or comprehensibility of the information, it is the psychological and mental way in which those who are still unwilling deal with climate information. For example, because they either see their self-image and their individual freedom threatened by environmental and climate protection measures, or because they avoid feelings of guilt and do not want to change their habits and world views, they (usually unconsciously) activate psychological defense mechanisms.[69] Environmental and climate psychology shows us that it is not so much education or knowledge – nor age, gender, origin or lifestyle – that determines climate skepticism, resistance, ignorance and inaction. Rather, a person's political orientation, beliefs and values, and social identity has a major bearing on whether they are opposed to climate protection.[70] In the words of Norwegian psychologist and climate communication expert Per Espen Stoknes: 'If there is a conflict between the facts and a person's values, the facts will lose.'[71]

But the good news is that we don't need those who are still unwilling – at least not necessarily. Because in our democracies, decision-making is based on the majority.

And fortunately, a clear majority of people now recognize the climate problem and are in favour of ambitious climate policy measures.[72] Even if the sceptics are sometimes quite loud or appear dominant, especially in social media, we must not forget that they are in the minority. And they won't be reading this book anyway. So instead of working my way through hard nuts and trying to get those who are still unwilling to change their minds, this book is aimed at the open-minded and interested people from mainstream society. I would therefore like to focus on the other three (stereo) types of the footprint matrix.[73]

'Unfortunately, no one can be [just] told what the Matrix is', says Neo's mentor in the movie, 'you have to see it for yourself.' That's exactly what we're going to do now. We will get to the root of the effects of the carbon footprint propaganda among the 'Haphazard Doers', the 'Internally Torn' and the 'Temporarily Frustrated'.

Chapter 3: The Haphazard Doers and their footprint errors

On a cold winter morning in 2016, a student in jeans and a dark blue winter jacket enters a supermarket in Germany. The night before, he watched the climate documentary *Before the Flood* by Leonardo DiCaprio and Fisher Stevens. The eerie images of collapsing icebergs in Greenland, flooded streets in Miami and Canadian oil sands fields are still buzzing around in his head. The warm fan at the entrance ruffles his hair, and the disturbing images fly away. What's on the shopping list? Oh yes: tomatoes, soy milk, meat and a new toothbrush.

It happens in the fruit and vegetable section immediately after the entrance: the tomato dilemma. A variety of different tomatoes from the region or Spain – including conventional cherry, cocktail and vine tomatoes, and all of them organic – are offered to the student in large and small containers made of cardboard or plastic. How should he decide? Normally he would simply choose according to his taste and look at the price tags, but after last night's shocking climate documentary? Impossible. He thinks he knows that organic tomatoes are more climate-friendly than the conventional ones next to them. But the organic tomatoes are shrink-wrapped in plastic, and also more expensive. He hesitantly reaches for the conventional vine tomatoes from the region without plastic – despite the queasy feeling that he has made the wrong choice in terms of climate friendliness.

He moves on to the plant milk section (which isn't actually called that because, by law, only the liquid from animal udders can be called 'milk'). At the shelf with soy, oat and almond drinks, the next dilemma looms: What was that again? Isn't the rainforest in the Amazon being cut down to grow soy? And the almond plantations in California use so much water? He resolutely reaches for the oat milk and makes his way to the meat department.

When he sees the shimmering reddish steaks and sausages behind the glass, he gets that queasy feeling again. This time, however, it's not because he's unsure after making a choice, but because he's already sure before he makes it. Because one thing is clear to him: meat is bad for the planet. But it tastes so good and is part of a balanced diet, the student thinks. So, at least choose the most 'climate-friendly' meat: and so a piece of organic beef from the region without plastic packaging ends up in his shopping basket.

In the hygiene products aisle, our protagonist is pleasantly surprised to discover a 'climate-positive' wooden toothbrush. Although it costs one euro more than the plastic one next to it, the good feeling of making a contribution to climate protection is worth the small surcharge.

Then disaster strikes at the checkout: the student has forgotten his cloth shopping bag at home! Dejected, the unlucky student grabs a paper bag and places it on the shopping conveyor belt under the judgmental gaze of the other people in line – at least that's how he feels. It beeps four times, he pays quickly, then 'Thank you, goodbye' and he's done. But not quite: as he loads the groceries into the trunk of his car, he thinks: 'Actually, I could have walked that distance. But it's only a small car. It doesn't consume that much. And I've just done some super climate-friendly shopping.' The slamming of the tailgate chases his thoughts away, the student gets behind the wheel and starts the engine. A relieved and satisfied sigh.

...

I was this student.

No clue about climate action

With so much discussion about climate action and decades of educational work by ministries and authorities, NGOs and environmental associations, companies and clubs, you would think that people are now well informed about climate-friendly consumption and the key levers of their carbon footprint. But unfortunately, the opposite is the case: in general, we have hardly any idea about meaningful climate action on a personal level – even many people who consider themselves to be quite climate-conscious.[74] And the collective confusion about the real CO_2 levers in everyday life is not just a British phenomenon. Surveys in other countries such as France, Germany, the UK and USA show very similar results: Take for example the international analysis of consumer behaviour in 28 countries conducted by the market research institute Ipsos in 2021. It found that, on average, people in almost all countries surveyed – whether in Europe or on other continents – are more likely to do small things such as recycling, separating waste, saving water or avoiding plastic packaging to protect the climate instead of plugging the major sources of CO_2 in their lives.[75]

Of course, we cannot conclude from this that the many years of intensive public education about CO_2 saving opportunities have so far achieved nothing. But we have now reached a saturation point where even more and more detailed educational books, blogs and podcasts with CO_2 saving tips are no longer of much use. (An economist would say: 'The marginal utility of CO_2 education is declining sharply.') The reason is that the issue of greenhouse gas emissions is becoming more and more complex and requires more and more decision-making effort the further you go into detail – for example, when it comes to the questions 'Organic tomatoes in plastic packaging or conventional tomatoes in a cardboard tray? Oat, soy or almond milk?'.

In addition, we consumers are already confronted with an almost unmanageable flood of selection criteria: In addition to price and taste, issues such as plastic, animal welfare, organic, water consumption, origin, Fairtrade, ingredients, nutrition, child labor, sugar content and a multitude of seals and labels need to be considered. Nowadays, without a degree from a university for conscious consumption, no one can really find their way through the product jungle without a guilty conscience. Even detailed 'CO2 maps' are no help in this situation, because the product jungle only becomes more opaque with the CO2 label and because dozens of other maps point in different directions. A study by German behavioural economist Frank Bilstein shows that people hardly change their behaviour when presented with detailed information on greenhouse gas saving options.[76]

The result is that **the footprint focus turns many people into haphazard doers who follow footprint misconceptions they have picked up somewhere in good faith, instead of really making a difference by increasing their handprint as citizens.** On the whole, the Haphazard Doers perceive themselves as relatively climate-conscious, but in reality they are merely performing small symbolic acts for the climate. They often even unconsciously cancel out their CO2 savings elsewhere.

A closer look at the footprint fallacies

In order to illustrate the complexity and thus the excessive demands placed on us and the Haphazard Doers by the carbon footprint focus, let us look at the results of an exemplary survey conducted by Kearney in Germany in 2019. I don't want to produce *another* 'How to reduce your carbon footprint' guide here, but show that focusing on our individual carbon footprints confuses so many people and is not the way to go.

The following figure shows in a circle around the keyword 'belief' what the respondents believe contributes most to saving greenhouse gases in their everyday lives. The large black clouds 'facts', on the other hand, indicate which measures *actually* make a significant contribution to their per capita emissions. Faith on the inside, facts on the outside.

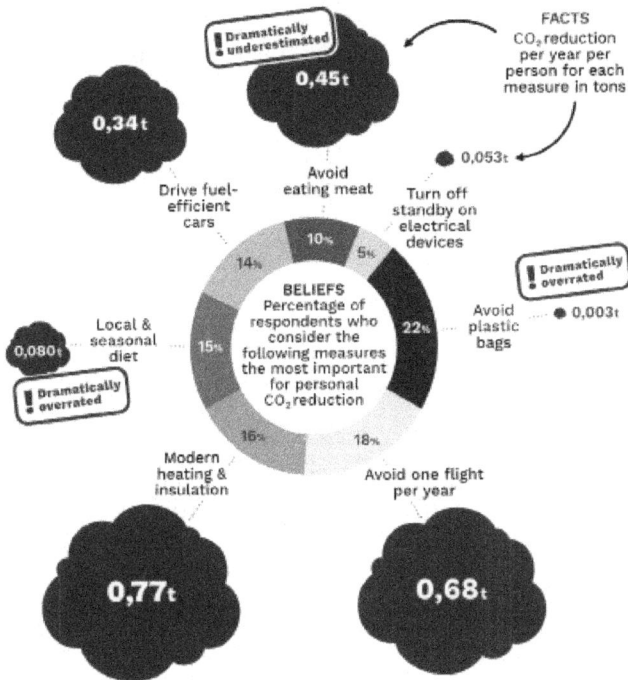

Figure 7: Belief vs. facts – results of the representative Kearney survey in Germany from October 2019 (the unit here means tons of greenhouse gases)

The three most glaring and common footprint errors are immediately apparent: many people overestimate the climate impact of plastic bags and a regional and seasonal diet. The negative effects of meat consumption are still greatly underestimated. In addition, beliefs and facts also differ to a greater or lesser extent on other topics.

Such misconceptions about reducing our carbon footprint can be discouraging. 'If I'm doing most of it wrong anyway, then I'd rather not do anything at all', some people think as their awareness increases, putting books like this one aside with resignation and distracting themselves. In doing so, however, they are merely changing the field within the footprint matrix from haphazard actionism to climate-conscious inaction, instead of overcoming the limits of the carbon footprint concept by adopting the handprint perspective.

EXPLANATION:
Single action bias and rebound effects

'The first principle is that you must not fool yourself – and you are the easiest person to fool,' the famous physicist Richard Feynman once said. Oh, how right he was. Even if we think we are always acting rationally and intelligently, we make blatant mistakes every day. In technical jargon, the psychological causes of such decision-making errors are called 'biases'. No one is completely immune to them, but we can at least become aware of them.

The single action bias: small symbolic acts to calm the conscience

The 'single action bias' is a particularly crucial distortion of thinking when it comes to personal climate action measures. For us humans, just one small or symbolic act is often enough to soothe our conscience and make a problem that would actually require larger or more action steps to solve seem less bad.[77] This one good deed is then often used as a compensation, excuse or apology for climate-damaging behaviour. For example: 'Yes, I drive an SUV, but at least I separate my trash at home' or 'If I'm going to cycle to the supermarket, I might as well buy meat'. In technical jargon, offsetting bad behaviour with small good deeds is called 'moral licensing'.[78]

If, for example, a friend who flies a lot asks me what is better for her carbon footprint – drying her hands with paper towels in the airport toilet or using an electric hand dryer – then the single action bias and moral licensing are probably at work. This is problematic in two ways: firstly, 'climate-correct' hand drying can of course not be an excuse or a conscience-soothing measure for a flight that is ten thousand times more harmful to the climate. And secondly, this kind of (inappropriate) self-calming and self-excuse stifles other, potentially bigger steps towards action. Because if you think you have already done enough with one small act, you will not do more – let alone enough.[79] In this way, the spiral of action fizzles out before a domino effect of many good deeds can develop.

The rebound effect: two steps forward, one step back

Even committed CO_2-savers are not safe from biases. In particular, widespread and insidious 'rebound effects' are throwing a spanner in the works for many climate activists. They mainly occur when efficiency improvements are made to save energy, time, money or CO_2. This is because such efficiency gains are often (partially) negated by increasing consumption or a change in behaviour elsewhere. This usually happens unconsciously, without us realising it.

One of the biggest time-saving rebound effects has probably been in communication in general. The proliferation of smartphones, emails, messenger services and social media means that we save an enormous amount of time per message to our acquaintances compared to traveling in person, sending a carrier pigeon or a fax or making a landline call. However, this time saving is being negated by the sharp increase in the number of people we are in contact with and the increased frequency of communication. Whereas a hundred years ago it took three hours a day for a face-to-face conversation, two detailed phone calls and three handwritten letters, today it takes three hours a day for ten emails, twenty text messages and thirty likes or comments on social media. As we can see: Despite communicating at the speed of light, there is (almost) no time saving because we communicate all the more and all the more frequently.

It works in a similar way when it comes to environmental protection. Firstly, environmentally harmful actions are often offset by environmentally friendly ones through moral licensing. This phenomenon can be seen as a psychological rebound effect. For example, a 2013 study found that people who reduce their water consumption because of a digital water meter suddenly consume more electricity.[80] Secondly, we humans tend to use more efficient technology more often, which reduces potential energy savings. For example, if you constantly switch on a new, super-efficient dishwasher half empty because it is so economical, you partially negate the potential electricity savings.

A number of scientists have already attempted to quantify the energy rebound effect. The range of their results varies considerably. For example, four overview studies came to the conclusion that direct rebound effects in the areas of transportation, buildings, households and electrical appliances in industrialized countries amount to roughly 10 to 30 percent.[81] Individual studies on the actual effect of more fuel-efficient cars even put the rebound effect here at 82 to 88 percent.[82] This means that if I were to buy a new car that on paper consumes five litres less fuel per 100km than the old one, then in practice I would actually only achieve fuel savings of 0.6 to 0.9 litres because I would unconsciously drive more often and differently. As the title of a rebound study carried out for the European Commission in 2012 puts it: 'Two steps forward, one step back'.

Super rebound effects: two steps forward, three steps back
But it gets even trickier: in special cases, it can even be 'two steps forward, three steps back'. Namely when we overcompensate for the savings, i.e. when there is a kind of 'super rebound effect'. In this case, our changed behaviour is ultimately even more harmful than if we had not attempted the savings in the first place. This was the case, for example, with a supermarket chain's 'Turn lights into flights' marketing campaign, which encouraged customers to buy energy-saving lights by offering them air miles for every light they purchased.[83] Of course, this campaign was quite counterproductive from an

environmental and climate point of view, because the CO_2 and energy consumption of the reward flights exceeded the savings of the energy-saving lamps at home many times over.

But the super rebound effect can also be less obvious. For example, if you spend the money you save from a lower fuel or electricity bill on even more consumption or an extra vacation flight. Or if you drive your SUV to an organic farm 10km away to buy regional and seasonal vegetables.[84]

Chapter 4: The Internally Torn and their guilty conscience

On a perfectly normal day, a student was sitting in a waiting room with two other people, doing what you do in a waiting room: waiting. Lost in thought, he picked up the top magazine on the pile of newspapers and began to leaf through it.

Suddenly it happened: smoke crept through the crack in the door into the waiting room. A fire must have broken out somewhere in the building, perhaps even in the corridor just behind the waiting room door. Startled out of his thoughts, the student reflexively looked up from the magazine to the other two people. They did not move and remained sitting impassively. The student was irritated by this and hesitated briefly as to whether he should jump up and do something, but then decided to remain seated, also outwardly calm.

The room became increasingly filled with smoke. By now it was obvious that the other two people must have noticed the smoke too. But they continued to sit calmly, as if everything was completely normal. Although the student was getting more nervous inside – after all, his life was in danger – he did the same and didn't let on. 'Stay cool,' he thought to himself, 'if it was really bad, the other two would do something.'

The climate bystander effect

That student was not me. But this scene really did happen, many, many times.[85] This is because it is part of an experiment from 1968 in which the two US social psychologists Bibb Latane and John Darley investigated the so-called 'bystander effect' in more detail.[86] This term describes the phenomenon that people generally tend to stand idly by in emergencies when other people are also aware of the emergency but do not react accordingly.[87]

What was proven by the above-mentioned smoke experiment from 1968 is astonishing: the bystander effect not only happens when other people are in distress, but also when you are in distress yourself. When the other people in the waiting room were instructed beforehand to remain inactive and apathetic despite the smoke, only 10 percent of the unsuspecting test subjects got up and did something. A full 90 percent, on the other hand, ignored the smoke and remained sitting impassively in the waiting room like the other two people.[88] Instead of rescuing themselves from the supposed danger to their lives, they sat idly by in a kind of self-inflicted bystander effect.

The smoke experiment can be applied to the climate crisis: Many voices in the climate movement repeatedly use the metaphor of a burning house to describe the climate emergency. And it is true: The earth is burning – literally due to the increasing number of forest fires. The smoke may not be creeping in through the cracks in most of our doors, but it can now be seen regularly on images taken by satellites from space. Almost all of us are in a similar predicament to the test subjects in the experiment described above due to the climate crisis.

Many people have realized this. They know that there is a fire. In other words, they see the smoke in the waiting room. But they still don't act. Instead, they fall into a kind of 'climate bystander effect'. I call this group the 'Internally Torn'.

Who are the Internally Torn?

But why are those who are torn inside not taking action despite being aware of the danger and having extensive climate knowledge? 'How do we sleep while our beds are burning?', sang the aptly named Australian rock band Midnight Oil back in 1987.[89] To answer this question, let's first take a closer look at who the Internally Torn are and how they are doing.

The Internally Torn have a greater climate awareness than the Haphazard Doers, and they are more familiar with the

CO_2 impact of various everyday decisions and behaviours. They therefore know that they cannot significantly reduce their carbon footprint with climate trifles such as cloth bags, wooden toothbrushes or insect hotels. No, they are aware that the central CO_2 levers mean a certain change in their everyday lives with a more or less high initial outlay. While the Haphazard Doers can not see the CO_2 jungle for the trees, so to speak, and naively run off with a good feeling, those who are torn inside know what a hardship the path through the jungle would be. And that leads them into a dilemma:

On the one hand, they want to reduce their carbon footprint. Because deep down, they have the desire and the moral requirement to be a good person who leaves a world worth living in for future generations. On the other hand, however, they do not want to accept the changes to the extent that they are necessary. So the problem is not that the internally torn do not know what they are doing, but that they do not do what they know. Against their better judgment, they fail in everyday life due to existing obstacles to climate-friendly behaviour. This creates a gap between what they actually want and what they actually do (i.e., the 'intention-behaviour gap').[90]

So even the path of seemingly least resistance comes at a cost, namely an emotional cost. Because the constant balancing act, the discrepancy between the inner moral claim and the lived reality, scratches the self-image of those who are inwardly torn. Whenever they get on a plane, order a steak in a restaurant or drive short distances by car and think about the climate crisis, feelings of guilt and shame arise. For short periods of time, distraction or appeasement help them. In the long term, however, this inner tension is quite stressful. (If you would like to find out more about the psychological background to inner turmoil, you can move forward to the explanation on cognitive dissonance and then return here).

External, structural obstacles

The first part of the answer to the question 'How do we sleep while our beds are burning?' is given by German climate activist Luisa Neubauer in her podcast: 'It is exhausting to find sustainability important in a world that is simply not sustainable.'[91]

As humanity has built an unsustainable world since industrialization, we have to make an enormous effort and incur costs in order to live in a climate-friendly way.[92] For the Internally Torn, these costs and efforts are just too high compared to the benefits of a significantly smaller footprint.[93] The social psychologists Andreas Diekmann and Peter Preisendörfer write that people will only act in a climate-friendly way if the changes 'do not require any drastic changes in behaviour, no major inconveniences [...] and no particular time expenditure'.[94]

A rocky day in the life of Julia

We are going to take a look at this using an imaginary daily routine which is peppered with many of my own experiences. The person we will be accompanying for a day as an example will be called Julia:

It's 7 a.m. and Julia's alarm clock rings. Yawning, she gets up and drags herself into the shower. The gas boiler starts up to provide her with hot water. The other day she met her landlord in the stairwell and asked him whether it would be possible to replace the old natural gas heating with a modern heat pump. But her landlord replied that this would not be possible. The reason, he said, was the preservation order, which forbids the necessary energy-efficient refurbishment of this old building.

As Julia prepares a freshly showered breakfast tea for herself and her daughter, her eyes fall on the kettle and she remembers the rest of the conversation with her landlord. The landlord had not only blamed the preservation order for the whereabouts of the natural gas boiler and the inadequate insulation, but also for the fact that he could not install a PV system on the roof. So instead of green solar power, the local energy mix including fossil gas is once again making the tea water boil this morning.

On the way to work, Julia drops her daughter off at the daycare centre by car. She would actually like to teach her daughter how to ride a bike, but unfortunately there is no safe cycle path from her home to the facility. And as Julia is provided with a company car by the company, including fuel costs, she drives the five kilometers to her workplace by car every day anyway – so she can take her daughter with her.

When she arrives at work, she immediately starts planning her next visit to the trade fair: she and her team from the purchasing department have to book their return travel from Newcastle to London for a trade fair. Most of the team vote to fly because of the time and money savings. Julia feels she has to join them because she doesn't want to be the only one who has to book a more expensive train ticket and leave earlier.

Lunch break. Julia's team members go to the canteen together. In addition to the standard salad bar, there are three dishes to choose from that day: Meatballs with potato salad, Shepherd's pie with beef mince and salmon with rice. Julia doesn't want to eat salad again and opts for the Meatballs. At the counter, she suddenly feels guilty because she had actually resolved to eat less meat because of greenhouse gas emissions. But the queue was long and the chef holds the steaming portion out to her urgently. She grabs it.

In the afternoon, she works through some orders and places orders for deliveries. She has to meet precise specifications and always choose the cheapest supplier – regardless of where the goods come from, how climate-friendly they were produced or how easy they are to repair, reuse or recycle at the end of their service life. She would prefer to buy only sustainably and regionally produced products and parts, but she is not allowed to do so. Although the company publicly claims to be particularly environmentally friendly, behind the green façade the motto is still: profit before planet and people, and Return on Investment before return and recycle.

After work, she quickly goes shopping while her husband picks up her daughter from the daycare centre. The last time she went shopping, she tried going to the organic food store. However, the bill was so high that this time she's going back to the cheaper discount store. On the drive home, a short feature on the climate comes on the radio: 'Steaks, SUVs and flying – are we out of our mind?' Julia gets a queasy feeling. She switches off the radio.

Climate-friendly options must become available, cheaper and more convenient

As we can see from Julia's daily routine, there are obstacles to climate-friendly behaviour everywhere in our everyday lives: In addition to the gas heating in the rented apartment, the electricity from burning fossil gas in the local electricity mix, the company car with an internal combustion engine, the lack of a cycle path or the meat-heavy canteen menus, there are also new buildings and highway expansions with the climate killers cement, steel and asphalt, the local diesel bus fleet and infrastructure designed for car traffic in cities, the school trip by plane or the financing of new oil and gas projects by banks and insurers. And so on and so forth. And in this world – with its smokestacks, roads, fossil fuel fired power plants, airports and beef production farms – Julia is now supposed to cycle more, use green electricity, book trains instead of planes and eat a vegetarian diet? Quite a lot to ask.

Fortunately, an increasing number of people are questioning the climate-damaging standards in our society and demanding alternatives. And indeed, such climate-friendly alternatives are thriving due to political measures and shifting demand. However, the climate-damaging standards are so heavily subsidized, so structurally entrenched and so deeply rooted in our culture that most climate-friendly alternatives are still lagging behind to a greater or lesser extent.[95] Roughly summarized, we can divide this lagging behind of climate-friendly alternatives into four categories (which can also overlap):

- Either the alternatives are widespread and therefore readily available, but **more expensive** than the standard option (Julia's organic store purchase).
- Or they are available, but **more time-consuming** (Julia's business trip from Newcastle to London).
- Or they are not so widespread, unknown or **do not exist** at all (the missing cycle path to the daycare center).
- Or climate-friendly alternatives are simply **forbidden** (the requirements of heritage and monument protection and the purchasing department).

This is the reason why only very few people can really live in a climate-friendly way within the current structures without making a significant additional effort. Most others continue to follow the 'normal' climate-damaging standards – including those who are inwardly torn.

In addition to the external, structural obstacles, the Internally Torn also face internal, interpersonal obstacles.

Social norms, the do-gooder problem & debates on guilt and shame

Most people want to behave according to the (unwritten) rules of their social groups. Psychology calls these rules, ways of thinking and behaviour 'social norms'.[96] Social norms are also present in the short story of Julia's daily routine. For example, Julia behaved in accordance with the norms towards her team when booking the trip to the trade fair in London: Although she would have preferred to book the train due to her climate awareness, she kept her mouth shut and followed the majority to avoid experiencing social rejection in the form of a condescending comment or a mocking remark.

Those who are inwardly torn therefore feel that a consistent commitment to climate action can put a strain on their social relationships. This is based on a general problem with visible moral behaviour, which I call the 'do-gooders' problem'.[97] Environmental psychologist Lise Jans and behavioural economist Jan Willem Bolderdijk aptly express this do-gooder problem in an academic article: 'When pioneers act out of moral concern, they may implicitly call into question the moral integrity and altruistic reputation of those who cling to the current, unsustainable status quo. As a result, members of the majority may react defensively [...].'[98]

Applied to private climate action efforts, this means that if people around those who are inwardly torn feel caught out or subtly attacked by their climate action attempts, they switch into defense mode. This happens above all when they identify strongly with eating meat, driving fast cars, owning a large house, traveling to distant holiday destinations or wearing the latest fashion trends. From their perspective, someone then seems to be attacking their identity and their positive self-image with moral justifications and unusual climate-friendly behaviour. As a result, the moral pioneers and do-gooders are typically belittled, ridiculed or subjected to mocking comments. This is often done subtly by disguising the criticism in the form of jokes. 'Save the world or keep your friends?' is how science journalist Christopher Schrader sums up this conflict of the Internally Torn.[99] For fear of social rejection or exclusion, many tend to opt for the latter.

The toxic and powerful mix of climate-damaging social norms, the do-gooder problem and the human fear of social rejection is joined by another obstacle: we still have far too many individualized guilt and shame debates when it comes to the climate issue. This is how bestselling author Frank Schätzing put it in an interview with me.[100] As feelings of guilt are highly unpleasant, we humans have a strong desire to avoid them. This is why the widespread accusatory culture of blame aimed at individuals is unattractive and repellent to anyone interested in climate and environmental protection. As a result, those who are inwardly torn engage less with their individual options for climate action. (This is why Julia switched off the radio on the way home after shopping.) If, on the other hand, the climate debate were conducted in an understanding, benevolent and – to counter the big word 'guilt' with an equally big word – *forgiving* way, moderate feelings of guilt could even become a driver for climate action.

Climate hypocrisy

The guilt and shame debates bring us to a related interpersonal obstacle: the fear of being accused of 'climate hypocrisy'. The underlying assumption here is that anyone who openly commits to the protection of our environment and the climate must first live a consistently climate-friendly life and have a minimal carbon footprint, otherwise he or she is a climate hypocrite.

Is this really the case? Let's start with a little thought experiment: Sina, Tim and Noura walk through a forest on three different days. First, Sina crosses the forest. She throws a to-go coffee cup between the trees at the side of the path. As she steps out of the forest, we ask her if she has thrown garbage into the forest on her way. Sina says sheepishly: 'No, I didn't.' The next day, it's Tim's turn. In his spare time, he is committed to environmental protection and takes part in public demonstrations against pollution. But he also carelessly throws his to-go coffee cup on the forest floor during the walk. Now the question: who do we find worse, Sina or Tim?

Most people probably intuitively choose Tim, who hypocritically campaigns for environmental protection and then throws garbage into the forest himself. In fact, psychological science shows that the majority of people find the Tim type, a hypocrite, worse than the Sina type, a liar.[101] The reason is that we humans deeply despise hypocrisy and hypocrisy because hypocritical behaviour can give us additional advantages at the expense of others. The loss of trust and social rejection of hypocrites is correspondingly high.

Now Noura comes into play. On the third day of our thought experiment, she strolls through the forest and also throws her to-go coffee cup at the side of the path. On the other side of the forest, we ask her if she has thrown garbage into the forest and she replies: 'Yes, I have. I admit it, I'm just not really into environmental stuff.'

Most people won't find Noura's behaviour okay, but they will probably find her somewhat likeable. Because at least she's honest compared to Sina and Tim.

The only problem is that the forest doesn't really care. In all three cases, the same amount of garbage was thrown into the forest. So all three people are doing exactly the same amount of damage to the environment. Rationally speaking, we should therefore find Tim, the hypocrite, the most likeable. This is because he is the only one of the three who is actively doing something about environmental pollution – for example, by campaigning for more garbage cans to be placed in the forest, for the public order office to monitor the forest paths and for the garbage to be collected in large-scale collection campaigns. If we really thought about the forest and the measurable effects, we would actually have to dislike Noura the most. Because she admits in full awareness and without a guilty conscience that she doesn't care about environmental protection and therefore throws garbage into the forest. Sina is at least aware of her shameful deed and lies shamefully.

Up to this point, I have copied this example of how we deal with hypocrisy from a science slam by German science journalist and TV moderator Mai Thi Nguyen-Kim.[102]

However, I am now extending this thought experiment to include a decisive factor in order to turn the example from the environment to the climate issue: How would our sympathy rating change if it were necessary for survival to throw plastic waste into the forest? What if it were impossible to cross the forest without leaving litter behind? Then it would be: Welcome to the climate crisis.

We have already discussed in this chapter that it is almost impossible to live in a perfectly sustainable and climate-friendly way within the current structures of our society. Everyone's life causes greenhouse gas emissions. In other words, every person leaves behind harmful waste when they walk through the forest. To put it simply, when it comes to the climate problem, there is also the Sina type, who lies and denies. There is Noura, who has nothing to do with climate protection and admits it publicly. And then there's Tim, who is publicly committed to climate action. (And there are of course many other types who take other positions, but it's not about them here). Although the Tim type helps the climate the most, although nobody can live without 'climate sin', most people despise supposed climate hypocrisy. (Jesus would probably have said: 'If anyone of you is without climate sin, let him throw a lump of coal.')

The thought experiment makes it clear that we are applying double standards in the climate crisis: The personal behaviour of people who speak out publicly about the climate crisis and call for climate policy measures is scrutinized closely. For everyone else, however, most people turn a blind eye (or both). The consequences are obvious: far fewer people find the courage to take a publicly visible stand for climate protection or even get involved because we still haven't collectively accepted that there can't be a perfectly sustainable life in an unsustainable world. 'Every week I speak to people from the public who say they are afraid to speak out about the climate because they expect a shitstorm because they [...] fly from time to time,' says German climate activist Luisa Neubauer in her climate podcast.[103] Due to our human fear of social rejection, accusations of hypocrisy are a huge obstacle that we put in each other's way, consciously or unconsciously.

Well-known 'climate hypocrites'

One prominent extreme example is the Hollywood star Leonardo DiCaprio. If you enter the words 'Leonardo DiCaprio hypocrite' into Ecosia or Google, you will get an idea of the outrage that is pouring out about his Janus-faced commitment to protecting the environment and the climate and his enormous carbon footprint at the same time. This was particularly evident in the months following the release of the Netflix film *Don't Look Up!*[104] (In the film, the imminent impact of a comet that threatens to wipe out all life on Earth serves as a metaphor for the impending climate catastrophe. Leonardo DiCaprio and Jennifer Lawrence play two scientists who together try to warn the government, media and public of the impending apocalypse. Unfortunately without success – and with a bitter ending). After the film's release, many voices attacked Leonardo DiCaprio personally. 'Don't look up ... you might see eco-hypocrite Leonardo DiCaprio on his £110 million yacht', read a headline in the Daily Mail in January 2022, for example.[105]

Yes, on the one hand Leonardo DiCaprio produces one of the largest carbon footprints in the world. Exact figures are not known, but private jet travel, luxury villas, cars and vacations on super yachts naturally add up to quite a lot on his personal CO_2 account. I wouldn't be surprised if Leonardo DiCaprio's carbon footprint is well over the 200 tons per year mark.[106] But on the other hand, he has been vehemently committed to environmental and climate protection for more than two decades, set up his own foundation for environmental protection, addressed the United Nations several times as a UN climate ambassador and published the influential climate documentary Before the Flood in 2016. He used his Oscar acceptance speech in front of over 34 million viewers to make an urgent appeal for more public pressure on politicians to finally take decisive action against the climate crisis.[107] This speech alone, just over a minute long, had a greater impact on public perception of the climate crisis than the 2015 Paris Climate Agreement negotiations or the annual Earth Day.[108] So Leonardo DiCaprio is working to break down the structural obstacles to climate-friendly behaviour at a political and societal level, so that the same people who call him a hypocrite can ultimately benefit and live more climate-friendly lives.

But it's not just celebrities with extra-large carbon footprints that are caught in the crossfire of hypocrisy accusations whenever they speak out about the state of the planet. Even people who keep their carbon footprint to a minimum are not safe from accusations of climate hypocrisy. Greta Thunberg, for example, was criticized in 2019 for crossing the Atlantic in a sailboat because the resulting media hype apparently caused more greenhouse gases than if she had simply flown.[109] (And just think: parts of the sailing boat were even made of plastic!) The German climate activist Luisa Neubauer was subjected to the hashtag #longdistanceluisa after climate change deniers spread old travel photos of her on the internet. And I, too, have often been accused of preaching the proverbial water while drinking wine – for example, when I stood up for climate action measures at a family party, having travelled there by car.

Why accusations of climate hypocrisy are mostly unfounded
However, all these accusations of climate hypocrisy have a crucial catch. The assumption underlying all accusations of hypocrisy is that you can only publicly demand what you yourself are already living. In general, and when it comes

to moral issues that are really in your hands, this is a very human, understandable and somehow logical point of view. With regard to the climate crisis, however, it would mean that only a vegan person who only wears second-hand clothes, never drives a car and never flies, who doesn't turn on the heating in winter and gets the little electricity they use from a mobile solar panel would be allowed to protest at climate demonstrations. (But woe betide her if she eats avocado toast while doing so!) Our human moral judgment is not well equipped to deal with such a collective, structurally anchored and complex problem as the climate crisis.[110]

If the structures in our society do not (yet) allow us to live a completely sustainable life, do we have to move into a self-sufficient wooden hut in the wilderness? No. If you can't and don't want to live this way because the obstacles are too great, you have to demand that these obstacles are resolved or reduced by those responsible in politics and big business. In other words, **we must first demand collectively what we cannot yet live individually, so that we can finally live what we demand.** Even if this runs counter to our usual moral compass. We don't need *less* climate hypocrisy, so to speak, but *more* of it.

However, the word 'hypocrisy' is actually wrong in this context. Because these supposed climate hypocrites are not criticizing the individual behaviour of their fellow human beings per se, but the structures that allow and encourage the climate-damaging behaviour of individuals. For example, anyone who calls for structural political measures such as a speed limit, a ban on natural gas heating or higher carbon pricing and then complies with these measures once they have been introduced is *not* a hypocrite. After all, this person drinks the water they preach as soon as it is available. Or is a Monopoly player a hypocrite just because she demands fairer game rules from the game manufacturer for the entire game community, which she would of course abide by after their introduction, but until then continues to play with the old game rules, just like everyone else does? No. Because she is not asking the individual players to (voluntarily) play differently, she is asking those responsible for Monopoly to change the rules. Applied to the climate crisis, this means demanding better political rules and framework conditions that prevent the global game of Monopoly from ending in disaster. From a purely logical point of view, the accusation of climate hypocrisy therefore collapses in most cases.[III]

The footprint focus as an amplifier of the obstacles

At the moment, there are still too many obstacles for the Internally Torn to become active in climate action. The basic problem is that focusing solely on the carbon footprint does almost nothing to help them overcome external, structural obstacles. Overall societal, structural problems such as the climate crisis are too big to be solved solely in one's private life and with a few altered consumer decisions. And to make matters worse, this focus on our individual footprints can even *increase* internal, interpersonal obstacles. It makes the path to sustainable living even rockier than it needs to be.

Despite all the hardships, only a small group of eco-heroes manage to tread this rocky path, overcome all obstacles and significantly reduce their carbon footprint: the 'Temporarily Frustrated'. While the costs of climate action were the decisive factor for the Internally Torn in this chapter, the benefits outweigh the costs for the Temporarily Frustrated in the next: a clear conscience, the image gain of a green lifestyle, possibly even a certain moral arrogance, but also positive side effects of a low carbon lifestyle – for example, slowing down, less stress and a healthier lifestyle through more exercise and less meat consumption. However, their heroic and exemplary behaviour takes its toll at some point, because the lifestyle of the Temporarily Frustrated can also be very stressful at times due to their excessive focus on their individual footprint.

EXPLANATION:
Cognitive dissonance

We humans have a need for what goes on in our heads to match our actions. In other words, our values, thoughts and behaviour must feel consistent. (In psychological jargon, we have a need for 'internal consistency of all cognitions'). If this is not the case, a 'cognitive dissonance' arises.[112] This is one of the central psychological phenomena when dealing emotionally with the climate crisis, because climate-conscious people are not (yet) able to behave in a completely sustainable way within the current structures of our society. Their thoughts, values and actions do not quite fit together. And this dissonance, distortion or tension can feel quite uncomfortable.

In my opinion, the comparison to music is very apt. Because the word 'dissonance' is made up of the Latin words 'dis' (apart) and 'sonare' (to sound). So it's like an orchestra in which half the instruments are tuned half a note too high. The state of cognitive dissonance is highly unpleasant, just as a symphony concert with out-of-tune instruments would be a highly unpleasant experience. This is why we – or more precisely our psyche – have a strong need to either suppress this dissonance through distraction or actively resolve or reduce it.

The most natural reaction is the former: suppress, distract, ignore. When the climate issue comes up and causes cognitive dissonance, you can simply turn your attention to another topic – for example, continue watching the TV series you've started, call a friend or put on your running shoes to clear your head. We can stuff earplugs into our ears in the concert hall, so to speak, and then pull out our smartphone to forget about the off-key music. The only problem is that when it comes to an issue as big, present and, above all, worsening over time as the climate crisis, this strategy is only helpful in the short term and seemingly so. Repression and distraction are not enough for most people because the cognitive dissonance is constantly resurfacing. Time and again, the climate crisis rips the earplugs out of our ears and the weird music becomes audible in the long run, there is no escape. Even with its head buried in the sand, the ostrich soon gets pretty hot at its buttocks.

Dissonance reduction

Instead of repeatedly trying to put earplugs in their ears and ignore the weird music, the psyche of many people also chooses the second active path. In psychology, this is called 'dissonance reduction'.[113] The aim is to create a little more harmony in the mental orchestra. But how does this work? The forefather of the theory of cognitive dissonance, social psychologist Leon Festinger, was already thinking about this. According to his research, our psyche can achieve dissonance reduction in two simple ways:[114]

- You can either change your *behaviour* so that it matches your beliefs. As I said, this option is very difficult in our current context.
- Or the psyche changes the *beliefs* so that they match the behaviour. Although this is the more convenient way, it is based on a mental distortion of reality or a psychological trick. The Unwilling, for example, fall for this trick: To avoid having to question their self-image or habits and laboriously change their behaviour, they are convinced that the climate crisis is not that bad after all or that they themselves or their country bear no responsibility. (This type of dissonance reduction even goes against scientific facts and, in extreme cases, can lead to belief in conspiracy theories).

This second option in particular – adapting your beliefs to your behaviour instead of the other way around – is very common when it comes to the climate issue. Sentences such as 'China has much higher CO2 emissions than the UK', 'The energy transition is too expensive' or 'First, we need more technological progress' help many people and their mental harmony, as they reduce the perceived need to become active and actually fundamentally change their behaviour. Most such sentences are not fundamentally wrong, which is why they are particularly suitable for adapting beliefs to the climate-damaging status quo of one's own life. In addition, the explanation of the single action bias shows that you can also reduce your cognitive dissonance with small symbolic individual acts: Excuses and mental backdoors such as 'I separate my garbage at home', 'I cycle to the bakery on Sundays' or 'My toothbrush is made from sustainable wood' can justify otherwise climate-damaging behaviour.

The point to remember about cognitive dissonance is this: we all want to bring harmony to our mental orchestra in the face of the existential threat posed by the climate crisis. In doing so, however, we should choose the one and only beneficial path: adapting our behaviour to our beliefs and values. Preferably with the handprint concept and without excuses, bogus arguments or mental backdoors.

Chapter 5: The Temporarily Frustrated and their feelings of powerlessness

'And then, we took out a friend of mine's speedboat and burned through fifty liters of fuel in just one hour. It felt like those things were going 150 miles per hour on the water – unbelievable!' the man in his sixties exclaims in an excited voice, his wide eyes glint with enthusiasm as he looks at Laura. They are standing opposite each other at a champagne reception in the garden of a finca on Mallorca. Laura had traveled for two days from London to Palma on four different trains and one ferry in order to be able to attend her aunt's birthday party, despite her decision not to fly anymore. The man, on the other hand, had flown over her head to Mallorca by plane – some forty hours after Laura had set off by train.

Laura is shocked by this fossil-hedonistic attack, but doesn't let on. The man continues with his unsympathetic small talk, talking about his newly purchased vacation home in Mallorca, including his new diesel SUV. Fortunately, before Laura's poker face begins to crumble, the man says goodbye and leaves through the crowd of guests. Laura is left confused and frustrated. The journey to Mallorca has been time-consuming and expensive. In the end, Laura has saved a few hundred kilograms of CO_2. But as long as people like this man – many people, thousands, probably millions – live so unconsciously and wastefully, there is no hope of curbing the climate crisis. Laura is certain of that. She sighs and breaks free from her torpor.

Laura's mood improves only marginally by dinner time. Although she has overcome the initial shock of this monologue, which has completely fallen out of time, the deep feeling of frustration remains as she sits with her family at one of the beer tables in the finca's garden.

'Is something bugging you?' asks her mother during the starter.

'Yes, an old white man who apparently thinks it's cool to brag to a younger person about how he can blow as much CO2 into the air as possible in the shortest possible time,' she replies snippily. Laura's father feels offended: 'Is the moaning and complaining about the global climate and the 'nasty old white men' going to start again? Can't you just enjoy a relaxing evening instead of constantly bringing up this doomsday mood?'

Suddenly Laura blurts out: 'No, Dad, I can't – especially not during a British birthday party in Mallorca, where some of your friends are saying things that just don't make sense in the mid-2020s! The Intergovernmental Panel on Climate Change, the UN Secretary-General and all serious scientists have been ringing the alarm bells for years. The world is heading for a deadly disaster of uncontrollable climate catastrophes and a sixth global mass extinction if we don't take radical action now.'

Laura now looks her father straight in the eye and gets louder: 'Have you ever really imagined what this means for me and my generation? The collapsing climate tipping points, the unstoppable rise in sea levels, widespread famine due to crop failure in the world's bread baskets, hundreds of millions of climate refugees, political systems destabilized by right-wing extremist parties, wars over fresh water between nuclear powers ... When I'm as old as you, large parts of the earth will be uninhabitable. And then this guy proudly tells me to my face how cool it is to burn fifty litres of fuel in an hour? And expects me to applaud?'

Laura notices tears welling up in her eyes. 'You know exactly how hard I work my ass off every day to preserve a future worth living for my generation. I only buy second-hand clothes, I'm vegan, I eat as regionally and seasonally as possible, I no longer fly, I don't have a car, I live in an energy-saving shared flat, I take cold showers in the morning and I avoid waste wherever I can. At some point, enough is enough. At some point it's over. At some point, I can't do this anymore ...'

At this point, Laura chokes up, and her voice breaks. All the pent-up, held back and overplayed emotions such as anger, disappointment and sadness begin to break out. To avoid the embarrassment of a crying fit in front of her parents, she jumps up and stormily leaves the table. Her mother and father sit there frozen. There is a heavy silence at the dining table, surrounded by scraps of conversation, clattering cutlery and buzzing insects. No one moves, not even Laura's siblings. After what feels like an eternity, Laura's mother breaks the silence: 'Who's going to join me for the main course at the buffet?'

Who are the Temporarily Frustrated?

Laura's story is exemplary for the group of the Temporarily Frustrated – the fourth and final field of the footprint matrix. A pronounced climate awareness gives Laura and all the other Temporarily Frustrated varying degrees of climate anxiety. They cope with these feelings of anxiety by actively trying to cause as few greenhouse gas emissions as possible in their private lives. This means that for this group of people, climate anxiety is generally functional, i.e. it activates and motivates them to act – except for a few brief phases in which it can become so strong that it turns into despair and has an inhibiting effect. (It might be a good idea to skip ahead to the excursus on climate anxiety and then return to this point).

The daily battle against structural and interpersonal obstacles

The Temporarily Frustrated spare no expense or effort to reduce their carbon footprint as much as possible despite structural obstacles. Heroic, consistent and disciplined, they pursue their mission of personal climate neutrality with zeal and vigor. Accordingly, they – like Laura – have already largely changed or reorganized their lives: They may have already got rid of their car, replaced their gas heating with a heat pump, switched their electricity tariff to one hundred percent green and taken meat off the menu. Instead of taking the plane, they travel by train, their clothes come from second-hand stores and the little waste they produce is, of course, properly separated.

To make matters worse, however, the Temporarily Frustrated also face headwinds from their social life. Laura's initial story illustrates this. Instead of receiving recognition, appreciation and support for their already arduous 'swimming against the current', the Temporarily Frustrated risk ridicule, conflict and exclusion.

It is therefore not surprising that only a few people genuinely strive to live significantly more climate-friendly lifestyles. While exact numbers for the UK are not readily available, the German Federal Environment Agency estimates that consistent climate protectors make up 14 percent of the population in Germany. Among them, the particularly serious CO_2 savers like Laura are an even smaller minority.[115] So while the Temporarily Frustrated sacrifice themselves heroically for the common good and their own conscience – sacrificing many personal comforts – they have to watch every day as the majority of their fellow human beings simply carry on as before. No wonder that frustration can arise.

The footprint sound barrier

But what are the benefits of their sacrifice and renunciation? First of all, we need to realize that even the greatest efforts to save CO_2 at home have their limits. The (infra-)structures of our society do not yet allow for a climate-neutral life. This is simply because we individually cannot directly influence the so-called 'public emissions' – for example for road construction, the military and the operation of schools, hospitals or street lighting. But these emissions still have an impact on our private CO_2 account. The self-experiment by German ZEIT journalist Petra Pinzler and her husband Günther Wessel demonstrates this particularly impressively. Together with their two children, they spent a year trying to live as 'carbon-neutrally as possible', as the subtitle of their 2020 book *Four for the Climate* puts it.[116] At the start of their self-experiment, the Pinzler-Wessel family's combined greenhouse gas emissions averaged 42 tons per year. Then they got started: they noted down small and large climate sins, researched and discussed alternatives and changed old habits. After a year of a strict CO_2 diet, the family of four took stock: 29 tons – a reduction of just 31 percent.

This result is an example of what many scientific studies on the topic of saving CO_2 in everyday life show: **even with real determination and a great deal of effort, average earners in an industrialized country like Germany or the United Kingdom will not realistically reduce their own per capita emissions by far more than a third.**[117]

Research suggests that in Western European countries, there is a kind of 'carbon footprint sound barrier' – depending on your country and living conditions at around 30 to 50 percent carbon footprint reduction – that we are currently unable to break (assuming we still want to lead a reasonably normal life).[118] The closer you get to this particular sound barrier with your private CO_2 savings, the stronger and greater the obstacles and resistance and the financial and time expenditure. I myself have been working for many years to significantly reduce my personal carbon footprint, but I am making similar experiences like the Pinzler-Wessel family.

Negligible effects

However, from a global bird's eye view, we can see how little it achieves to reduce per capita emissions as much as possible, i.e. to struggle away at our footprint sound barrier: even if a person with an average annual carbon footprint of 5 tons would somehow manage to reduce it to zero (which is impossible in today's infrastructure), the effect would evaporate as a small drop of CO_2 in the ocean of air. Humanity emits around 50 *billion* tons of greenhouse gases every year. (For comparison: this corresponds to the weight of around 135,000 Empire State Buildings – or roughly the weight of Manhattan in New York City). Trying to significantly reduce this gigantic amount of emissions with a saving of 5 tons is like scooping five bathtubs of water from Loch Ness. The fact is: It would only make a difference of 0.00000001 percent.

As a child, I often heard my grandmother say, 'Many a little makes a mickle' – but 0.00000001 percent is extremely little. My professors in mechanical engineering called such amounts *negligible*. On the other hand, large livestock make a lot of muck. Similarly, the major contributors to the climate crisis are the 'big cattle' – hidden within the global web of big business and neoliberal, old-economy-conservative or right-wing politics. For example, just one hundred corporations are responsible for over 70 percent of all global industry-related greenhouse gas emissions since 1988, and have thus caused around half of the earth's overheating to date.[119] The big oil companies are of course at the forefront of that list.[120]

The lockdowns at the start of the coronavirus pandemic in spring 2020 made it clear that we can only move this big cattle slightly in total, even with severe restrictions on consumption worldwide: When hundreds of millions of people around the world were no longer able to fly, drive and consume as usual, global CO_2 emissions only fell by around 6 percent.[121] Nevertheless, this was the biggest CO_2 reduction in human history. But firstly, despite large individual CO_2 savings by so many people, a full 94 percent of the CO_2 problem still remained. And secondly, this was only a one-off saving effect.

After the easing of the restricting measures, CO_2 emissions jumped back to pre-coronavirus levels – and even climbed to a depressing all-time high in 2024.[122]

The more clearly the Temporarily Frustrated recognize the imbalance between their individual sacrifice and the tiny overall effect, the more frustrated they become. Some of them are even quietly and secretly gnawed by envy of a carefree (consumer) life on a large carbon foot. 'Ignorance is bliss', they whisper.

In my opinion, the opposite is the case: we will only achieve bliss as humanity if as many people as possible understand their unpleasant but appropriate climate feelings as motivational energy for action and direct it primarily towards their climate handprints.

A little frustration is okay – and even helpful!

Internationally, more than half of 16 to 25-year-olds feel sadness, worry, fear, anger and powerlessness as their main emotional response to our situation in the climate crisis.[123] From surveys like this, we can conclude that most teenagers and young adults are more or less frustrated when they think about the climate crisis. No wonder: anyone who perceives the destructive climate consequences of the current level of global warming of around 1.3°C, thinks of the threat of even more devastating climate disasters in the future and at the same time observes the inadequate action of our governments, automatically feels a certain degree of frustration. This is a natural and appropriate emotional reaction to our situation. It is the price of knowledge – but also of the solution.

Natural grief and solastalgia

For example, it is normal to feel sadness in the face of burning rainforests, collapsing glaciers, dwindling sea ice, parched land and deadly floods. Such grief over the loss, change or destruction of one's own habitat is called 'solastalgia' in psychology.[124] For example, when I stand on one of the observation towers at the Hambach opencast mine near my home town of Cologne in Germany and look down into the dark coal crater, which is almost 400 metres deep and stretches to the horizon, I feel this kind of grief called solastalgia. I think of the land of Mordor from *The Lord of the Rings* and my throat tightens. I feel like crying. It's no wonder that residents living near such open-cast mines have above-average rates of mental and emotional stress and depression.[125]

I had a similar feeling when I saw the images from the Ahr Valley in the Eifel region, which I know well, after severe flooding devastated the region in July 2021 and claimed over 200 lives. The feeling of solastalgia also comes to me when I think of the scene with the walruses from the documentary *Our Planet*, narrated by Sir David Attenborough: thousands of walruses save themselves to rest on a much too small, rocky stretch of land because, due to global heating, there is no sea ice far and wide where they normally rest after hunting in the Arctic Ocean. As walruses are not made for climbing on rocky ground, dozens of them fall off the cliffs to an agonizing death in front of the horrified camera crew.

Yes, grief in light of what is happening to our ecosystems is an appropriate emotional response – and should be normalized – if one is capable of feeling compassion for the suffering of fellow living beings and humans.

Natural anger and disappointment

It is also normal that Laura's first reaction to the fossil-hedonistic speedboat bragging is anger, and that she is later disappointed by her father's reaction at dinner. The unsympathetic behaviour of the two men violates her moral convictions. Anger is a natural human reaction, especially to injustice. Laura is no different to the founder of the Potsdam Institute for Climate Impact Research, Prof. Dr. Dr. Hans-Joachim Schellnhuber, one of the world's most renowned climate researchers. In an interview for the Climaware podcast, I asked him how he feels about the climate crisis, and he replied: 'Sometimes I feel pretty bad. Then there are days when I feel better. [...] The overriding feeling now is actually anger at human stupidity – and I include myself in that.'[126]

I find the anger and frustration of many climate scientists deeply understandable. For decades, they have been warning governments, the media and the public about the increasingly catastrophic consequences of the heating of our planet and the disbalancing of our ecosystems. But for the most part, they have been met with rejection, deaf ears or mere lip service.[127] Like Hans-Joachim Schellnhuber, the Secretary-General of the United Nations, Antonio Guterres, now makes no secret of his anger. On the occasion of the publication of the last part of the latest major IPCC report, he delivered an emotional incendiary speech: 'This report [...] is a litany of broken climate promises. It is a file of shame, cataloguing the empty pledges that put us firmly on track towards an unlivable world.'[128]

Anyone who hears such statements from the head of the UN, like Laura, who still has most of her life ahead of her, but is not understood or taken seriously by older adults, can understandably feel anger and disappointment. This is also a result of the aforementioned international youth study from 2021: 16 to 25-year-olds who raise the climate issue and their concerns with their parents or other caregivers are usually ignored, dismissed or, at best, placated – just like Laura in the introductory story.

Climate emotions can be functional

But now for the good news: a certain amount of frustration is not only understandable and okay, in moderation frustration can even be helpful, constructive and functional – just like climate anxiety. Ultimately, it was this kind of 'functional frustration' that led Greta Thunberg to the Swedish parliament instead of school on Fridays in the summer of 2018. Just one year later, millions of people took to the streets of Europe under the banner of Fridays For Future, while the then 16-year-old expressed her grief, anger and disappointment to the United Nations in a sensational speech.[129] Her example shows how functional and impactful moderate climate frustration can be.

Numerous studies show that, on the one hand, 'positive emotions' – such as interest, pride and hope – play an important role in collective climate protection actions, protests and support for climate policy.[130] On the other hand, 'negative emotions' are also very important – especially so-called 'group-based anger'.[131] For example, a 2022 study found that people engage in vegan activism more often when they feel angry – one of the dominant emotions in a state of frustration.[132] Moderate frustration with the status quo can therefore motivate people to act, politicize them and drive them to take to the streets to stand up for their personal values and beliefs.

Too much frustration is harmful

Frustration only becomes a hindrance when it becomes too great – or when we artificially make it greater than it actually needs to be. When frustration gets out of hand, it has a paralyzing and destructive effect. As with climate anxiety, there is also a pathological facet to climate frustration. In extreme cases, it can lead to complete despair and even become dangerous to life and limb.

Probably the most dramatic examples of such an extreme level of frustration are the self-immolations of two environmental activists in the US to draw attention to man-made environmental disasters and the climate crisis – one of them 60-year-old David Buckel in a New York park in 2018, the other 50-year-old Wynn Alan Bruce in front of the US Supreme Court in Washington DC in April 2022.[133] Less dramatic examples are the hunger strikes by different groups of climate activists in Vienna and Berlin in the period from 2021 to 2024.[134] These actions show what is possible at the end of the spectrum of frustration: health-risking protests and even suicide.[135]

For most people, however, extreme frustration leads in the long term to them either becoming increasingly radicalized and wanting to use property damage or even violence – or they give up at a certain point and assume that everything is hopeless and lost anyway. I personally cannot discern any significant violent radicalization within the climate movement. (Far more, I perceive violent fantasies emerging from the growing right-wing spectrum of society.) The biggest and most worrying trend in society as a whole is the latter: an increasing turning away from the climate issue, retreat into private life, distraction through consumption, and hopelessly shutting the eyes to reality.

Climate doomism: it's already too late anyway and everything is lost!

What is particularly risky about excessive and prolonged frustration is that this emotional state can lead to a paralyzing climate depression in the long term.[136] In this state, you have lost all hope, see only black and assume that every effort and action is in vain. In addition to your own zest for life, your motivation to act also drops to a minimum: 'Why fight when everything is lost anyway?' Studies show that the very pessimistic part of the European population believes particularly *strongly* in the climate crisis, but is remarkably *less* committed to finding a solution – only slightly more than the climate sceptics or the Unwilling.[137] Climate depression and doomism therefore lead to a state of inaction.

But beware: under certain circumstances, doomism is a mental trap set by our psyche. The opinion that it is too late anyway offers extremely frustrated people welcome emotional relief. After all, those who come to believe that it is already too late, climate protection efforts are hopeless and everything is doomed can just as easily return to the simple, careless old life and behave like the majority of other people. Saving CO_2 and all the sacrifices and efforts involved don't make any difference anyway – that's the mindset.

This psychological trap provides an effective starting point for powerful interest groups who want to deliberately thwart climate protection efforts. For example, Michael Evan Mann, one of the most renowned climate researchers in the United States, describes the phenomenon of climate doomism in his book *The New Climate War* not only as a highly dangerous phenomenon that makes social climate action more difficult. According to him, the phenomenon also provides substance for targeted diversionary maneuvers by oil and gas companies, their stakeholders and lobbyists – for example in the form of lurid and exaggerated doomsday articles or with the help of trolls and bots on social media.[138] Whether it's denial, doubt, technical pseudo-solutions or climate doomism – the fossil fuel and climate deflection lobby will use any means to drive people away from climate action or drive wedges into the climate movement.[139]

Why the carbon footprint does more harm than good

As we can see, the saying 'the dose makes the poison' applies to both climate anxiety and frustration. However, my experience is that the focus on the personal carbon footprint is a catalyst for excessive climate frustration because it creates an unhealthy and ultimately unhelpful negative spiral. It is based on the myth that everyone only needs to drastically reduce their individual carbon footprint through climate-conscious consumption and everyday decisions in order to solve the climate crisis. But the fight against the footprint sound barrier and the tiny effect seen globally prove that this solution strategy is not enough, or to be more precise, is unrealistic and wrong.

William Rees, the inventor of the Footprint, has also

recognized this: 'You can change your purchasing habits a little, but on the whole what we do as individuals is relatively trivial, because the heavy lifting – the kinds of things that would make a real difference – are actions taken for the common good.'[140] Even the supposedly exemplary Temporarily Frustrated are subject to this vicious circle: the more vehemently Laura and the others try to get their personal footprint down, the greater their efforts become. Since such self-mortification is not particularly attractive to the Internally Torn and the rest of the population, only very few people follow the example of Laura and the Temporarily Frustrated. However, if there is no imitation, it seems to the Temporarily Frustrated that their personal sacrifices are not bearing fruit. Because they do not recognize the effect of their emphatically climate-friendly behaviour in their social environment, they eventually believe that all their effort is completely in vain. The perceived imbalance between their renunciation and the 'business as usual' attitude of the majority of their fellow human beings increases their climate frustration enormously.

Collective self-efficacy is the key
The handprint perspective offers an effective means of breaking this spiral of guilt, powerlessness and frustration. Firstly, it can relieve us of exaggerated individual feelings of guilt. Because the question 'How big is your handprint?' bypasses the bad climate conscience and the usual guilt and shame debates about driving the car, eating animal products and flying. In my experience, most people feel called, relevant, motivated and inspired when they learn about their potentially large climate handprint.

Secondly, the handprint concept can free us (at least in part) from our perceived powerlessness in connection with the carbon footprint. The key to counteracting the low personal expectation of effectiveness is the feeling of *collective self-efficacy*. Research results confirm that feelings of powerlessness and perceptions of helplessness can increase through individualization and decrease through 'we-feelings' in a group.[141] Instead of dwelling on the carbon footprint of individuals, we should therefore always emphasize the collective possibilities of climate action and realize that we are not alone.[142] The German Astrophysicist and TV presenter Harald Lesch, echoing the philosophical publicist Hannah Arendt, told me: 'I always assume that I can't save the world on my own [...], [but] when people come together, you can expect miracles.'[143]

For example, I always feel how incredibly powerful, unifying and miraculous this feeling of collective self-efficacy in a large group can be when I march through the streets with tens of thousands of people in the global climate strikes. Then my feelings of powerlessness and frustration dissolve in the crowd. Incidentally, the reduction of feelings of powerlessness and helplessness seems to work best with a global sense of togetherness – as a globally feeling and thinking 'earthling' or 'global citizen'.[144] It is probably precisely this effect that I perceive so clearly in the global climate demonstrations.

And thirdly, the handprint can break the spiral of frustration because the feeling of togetherness actually achieves more for climate protection. A 2021 study confirms that people with a collective, policy-oriented sense of self-efficacy are much more likely to support ambitious climate policies, get politically involved or take part in climate demonstrations.[145] And this pressure on politicians can have the effect of changing the overall political and social framework conditions in favor of sustainability. With the climate handprint, we finally have a tool that does justice to the scale of the problem. (In psychology, this is referred to as an 'epistemic fit'.)[146]

The 'We together against the fossil world' mode is therefore balm for the frustrated soul. Ultimately, it can even set a positive spiral in motion that leads to a general change of perspective: from the individual to the collective, from independent and detached individuals to the co-dependent and connected community of a group or society, from passive consumers of the economic system to active citizens of democracy, from an inanimate environment to be exploited to an equally shared, living and enchanted world, from guilt, powerlessness and frustration to responsibility, duty and compassion.[147]

On guilt and responsibility in the climate crisis

At the end of this chapter, it is important for me to point out the difference between guilt and responsibility. For example, it would be far too harsh to find Julia from the last chapter guilty if she flies to the trade fair with her team, takes her child to nursery by car or eats meat in the canteen every now and then. Because you are only guilty if you could have acted differently. Having the choice between right and wrong is a basic prerequisite for guilt. But Julia doesn't really have a free choice. Her behaviour is largely determined by the climate-damaging structures and social norms around her.

However, this does not release her from a certain degree of responsibility to actively work towards changing these climate-damaging norms and structures.

In fact, *everyone* has a certain responsibility to play an active role in solving the climate crisis. But the degree of responsibility varies from person to person because it depends on their degree of freedom, abilities and capacities. At the United Nations level, this is expressed in the principle of 'common but differentiated responsibility'.[148] I think we should also take this principle to heart at a personal level. This means that each person only has to bear as much responsibility for solving the climate crisis as they are able to bear in view of their capabilities – or their external, internal or interpersonal obstacles.

The special responsibility of the rich

In most cases, this personal responsibility is linked to financial wealth. This is because more money means more opportunities, resources, freedom of choice and therefore less dependence on climate-damaging norms and structures. In addition, wealthy people, with their capital, contacts and professional (leadership) positions, have a correspondingly large influence on exactly these social norms and structures that shape the behaviour of all people – such as Julia.[149] In short, wealthy individuals are more resilient and possess significantly greater leverage to influence the workings of society compared to those from middle or lower income classes. They therefore also bear the biggest share of responsibility.

What's more, financial wealth usually goes hand in hand with particularly high per capita emissions. Firstly, internationally: the wealthiest 10 percent of the world's population have caused half of all greenhouse gas emissions since 1990.[150] And the per capita emissions of the richest 1 percent – which includes all people with a six-figure annual income – amount to no less than one hundred tons per year, i.e. over 70 times more than the 'poorer' half of humanity.[151]
But the greenhouse gas emissions of the rich are not falling. No, they have significantly increased their per capita emissions since 1990 and will probably continue to do so.[152] By contrast, emissions reductions in the EU, for example, are being achieved almost exclusively by *middle*-income groups.[153] This is due to the fact that 'climate policies such as carbon taxes have often disproportionately burdened low and middle income groups, while the consumption habits of the wealthiest groups have remained unchanged', writes French economist Lucas Chancel in the *World Inequality Report 2022*.[154] Policymakers should therefore focus much more on the emissions of the high-income classes.[155]

As we can see: the ball is primarily in the court of the rich (or rather our governments to introduce socio-ecological tax policies). However, most of them seem to be either unaware of their special responsibility or dismiss it with excuses, diversionary arguments, humor or irony. This was the result of a sociological study in 2023 on the typical justification patterns of wealthy people with exceptionally high carbon footprints. 'People talked about their lifestyle as if it were completely normal and as if everyone did it,' reports one of the authors of this study.[156]

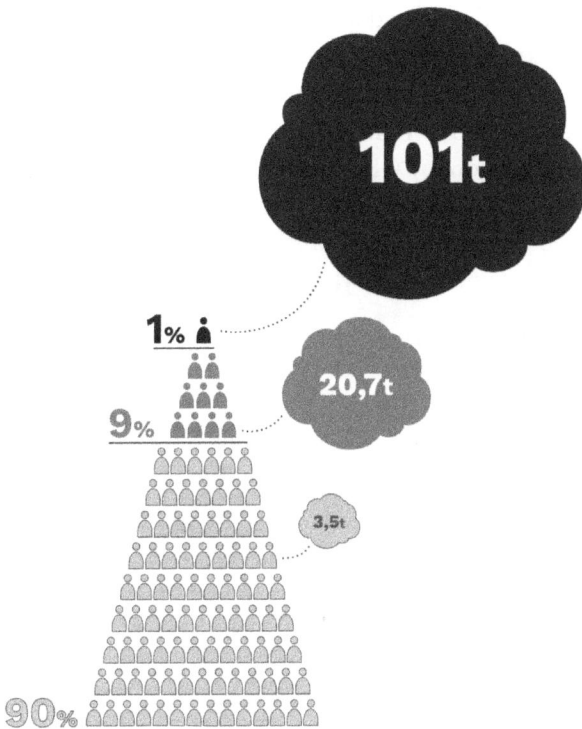

Figure 11: Global inequality of per capita greenhouse gas emissions (in tCO2 eq. and taking into account consumption, investments and emissions "hidden" in imports)

The current disparities in wealth, emissions and responsibility are not only deeply unjust: those who suffer the most from the climate crisis worldwide have generally contributed the least to it and have the least means to protect themselves or adapt.[157] What's more, the enormous inequality is downright poisonous for social acceptance of the climate transition.[158] Leading scientists are increasingly warning that it will be difficult or even impossible to get a grip on the climate crisis if the gap between the rich and the rest continues to widen – or the abundance of the few continues to increase at the expense of the many.[159] This is why so many climate activists are calling 'What do we want? ... Climate justice!' – between all generations, countries and income classes. I often think of a quote by Ghandi that I looked at almost every day for several months in the entrance area of the UN Climate Change Secretariat: 'There is enough for everybody's need, but not enough for everybody's greed.'

EXPLANATION:
Climate anxiety

Anyone who closely examines the scale and implications of the climate crisis is usually overcome with fear. I myself have experienced bigger as well as smaller waves of anxiety due to the beginning climate chaos.[160]

In psychology, there is now an established term for this: 'climate anxiety'.[161] And the phenomenon is on the rise worldwide – particularly among young people: 60 percent of 16 to 25-year-olds are 'very' or 'extremely' concerned about the climate crisis.[162] Sadness, worry and fear, but also anger and powerlessness are the dominant feelings of the younger generation when they think about the climate crisis.[163] Around 40 percent of respondents are even hesitant to have children because of the threatening climate prospects.[164] In line with this, search queries for 'climate anxiety' on Google and social media portals have been increasing significantly for several years.[165]

Fortunately, the number of books, podcasts, blogs and self-help guides on the subject is also increasing at the same time.[166] (The chatbot ChatGPT from OpenAI, for example, advised me to inform myself, take action, connect with others, practise mindfulness and look at the positives of the climate transition). And psychological experts are also reacting accordingly.[167] The Psychologists and Psychotherapists For Future (from here on I will use the abbreviation Psychologists For Future) in particular have taken this to heart and have been actively working for years to educate society about their guiding principle: 'Climate change is a *psychological* crisis – whatever else it is.'[168]

However, it is important to emphasize that anyone who feels climate anxiety does not have a pathological anxiety disorder. Lea Dohm, a German psychologist and co-founder of Psychologists For Future, emphasized this to me in our podcast interview.[169] So you're not mentally ill just because you're afraid of the consequences of the climate crisis. Quite the opposite: fear is a normal emotional reaction to the global climate emergency. What's more, in moderation, consciously perceived fear can even be beneficial and helpful. This is because fear is an important emotional indicator that can motivate us to take action. My climate anxiety, for example, played a key role in my decision to join climate protests, start my own climate podcast 'Climaware' and ultimately to write this book.

However, the two points 'in moderation' and 'consciously perceived' should be emphasized. For those who experience *excessive* climate anxiety or *none at all* (due to distraction and repression), climate anxiety is not motivating, beneficial or helpful. The three members of Psychologists For Future Felix Peter, Katharina van Bronswijk and Bianca Rodenstein describe that these two factors – the extent of the feelings of anxiety and their conscious perception – result in three facets of climate anxiety.[170] These three facets of climate anxiety are based on the three typical ways in which humans respond to danger: fight, flight or freeze.

The three facets of climate anxiety: functional, distracting, pathological

The first of these three facets of climate anxiety is functional anxiety ('fight'), i.e. moderate anxiety that is clearly perceived instead of suppressed by the person concerned. Coupled with options for solutions and courses of action, this fear can have an activating effect and motivate us to fight for our well-being and for the protection of our planet.[171] It is the kind of fear that Greta Thunberg meant when she proclaimed at the World Economic Forum in Davos in 2019: 'I want you to panic. I want you to feel the fear that I feel every day. And then I want you to act.'[172] Admittedly, the word 'panic' is somewhat unfortunate here, as panic usually makes you headless and can lead to irrational behaviour. But in the context of the speech ('And then I want you to act'), it becomes clear that Greta Thunberg meant *functional* fear.

The second facet of climate anxiety, on the other hand, manifests as *distraction* ('flight'). This coping mechanism operates in a similar way to distracting yourself from cognitive dissonance by putting earplugs in your ears in the audience of an orchestra that sounds out of tune. 'Many people completely escape from this fear without realizing that it is fear driving them to disengage from the climate crisis,' explains the German climate psychologist Janna Hoppmann.[173] While this avoidance may feel relieving for the individual in the short term, it poses a significant danger to society in the long term. Burying one's head in the sand – the ostrich tactic – is neither a solution to the climate crisis nor a sustainable way to cope with climate anxiety.

The third facet is the *pathological* form of climate anxiety. In this state, you are frozen, almost rigid with fear ('freeze'). Although the fear is not suppressed, it is not functional either. Rather, it has a hindering, paralyzing and inhibiting effect. Particularly on lonely evenings, when the distractions of everyday life subside and the body comes to rest, awareness of the global climate emergency can evoke very worrying and anxious feelings and thoughts that are acutely unmanageable. But even during the day, a very anxious mind requires a considerable amount of your own strength, vital energy and mental capacity – just as a running background program on your laptop constantly eats up working memory. I know this all too well myself.

This pathological facet of climate anxiety can be exacerbated when those affected can no longer sleep properly, refuse to eat or are unwilling or unable to leave their room. This was the case, for example, for the then 13-year-old daughter of a family friend from the Eifel region in Western Germany after a climate-induced flood disaster devastated the Ahr valley and her home region in the summer of 2021. As a result, her fear of heavy rain, storms, forest fires and other extreme weather events became so strong that she did not want to leave the house at times, had difficulty sleeping and lost all zest for life at certain moments.

Fortunately, there are also effective ways to find a good, functional way of dealing with climate anxiety, including open exchange with climate-conscious, understanding and empathetic people, active action and experiencing one's own self-efficacy in a sense of togetherness, and intensive nature experiences.[174] And, of course, professional help if necessary, for example from climate-conscious psychotherapists such as Psychologists For Future.[175]

Chapter 6: How to do it better: Making Footsteps with Big Points

Despite all the criticism, we should not completely throw the carbon footprint concept overboard because there are a few advantages to the concept. Used correctly and in addition to the handprint, reducing one's carbon footprint can even have positive effects – personally on our mental well-being as well as collectively on the climate transition.

A self-effective starting point

Most people begin their climate commitment with personal lifestyle changes, adjusting their individual behaviour. These small steps can lead to increasingly impactful follow-up actions. I experienced this myself: my own climate journey started with efforts to reduce my carbon footprint and eventually led me to full-time professional eco-activism. This progression is explained by what experts call 'positive spillover': when small, everyday actions – like sorting waste, buying second-hand clothes, lowering the heating, or having a few meat-free days per week – make it more likely that someone will take further climate-friendly actions.[176] However, for positive spillover to work, two key factors are crucial: acting in alignment with your own environmental values and allowing yourself to feel good (or even a bit of pride) afterward – without falling into the trap of single-action bias.[177]

Those who start their personal climate journey with their own behaviour experience immediate self-efficacy. Most of the changes that are necessary to reduce your individual carbon footprint can be felt directly in your own life: The light from the energy-saving LED lights in the office and switching them off after work, for example, can be seen with your own eyes (whereas the PV system on the company roof is rarely seen). You can feel the airstream, the powerful pumping of your legs and the light film of sweat on your skin as you cycle home (the petition for wider cycle paths, on the other hand, is not directly perceptible). And the vacation trip on the night train is an exciting adventure (although the carbon tax on intra-European flights is not). This direct visibility and experience of footprint actions can give us immediate empowerment and gratification: finally, we are no longer passive objects of climate impacts and policies, but active subjects for self-determined climate action in our own environment.

More self-sympathy and quality of life

When German sociologist Harald Welzer explained in an interview with the Swiss broadcaster SRF in October 2021 that he had quit smoking because he no longer found himself likeable ('sympatisch') as a smoker, something clicked for me.[178] I found the idea of self-likability in relation to one's own actions brilliant – because it makes the issue instantly clear and personal. It prompts a simple yet powerful question: Do I like myself when I fly to Italy for the weekend? Do I like myself when I eat meat almost every day? Do I like myself when I drive short distances in the city center?

Anyone who has broken – or at least weakened – the spiral of guilt, powerlessness, and frustration through the handprint perspective and, despite the tiny global impact, takes steps to reduce their own carbon footprint simply because it is the right thing to do, is likely to develop a greater sense of self-likability. This is because their external actions increasingly align with their inner values and beliefs (see *Explanation of Cognitive Dissonance*), leading to a greater sense of mental harmony.

In addition to increasing our handprint, reducing our

carbon footprint can even measurably improve our quality of life and overall well-being. In 2021, researchers were able to prove that almost 80 percent of all CO_2 reduction measures investigated by consumers had a positive impact on their well-being.[179] It is clear that eating less meat, cycling more and eating more organic food leads to better health.

Moreover, once material prosperity reaches a certain level, stepping out of the consumption-driven lifestyle imposed by advertising and current capitalist norms allows individuals to escape the social hamster wheel of excessive stress and performance pressure. Consuming less means needing less money; spending less money means working fewer hours; and working less for wages opens the door to a new kind of prosperity: time prosperity. This gives people more time for meaningful activities – work of the heart, fulfilling social engagement, hobbies, art, family, friends, love life, and more.

The power of consumers – every little counts (a little)

Small, individual, voluntary contributions at local level are not just 'nice to have' for a good feeling. No, they are important for the simple reason that we will not achieve the societal climate transition without behaviour changes: For example, in 2022, the UK Government's Climate Change Committee (CCC) calculated that over 60 percent of the necessary emissions reductions over the next years will involve personal behavioural changes.[180] Although climate-friendly technologies are also necessary for the majority of these, technical innovations alone will not be enough, as a large number of scientific reports have shown.[181]

The fact that we, as consumers, can influence, accelerate, and shape change to a certain extent is evident in the current rise of 'herbivores.' Today, almost every supermarket, canteen, bistro, café, and restaurant offers tasty and affordable vegetarian or vegan alternatives to meat, cow's milk, and other animal products. This shift hasn't been driven by policy changes, meat taxes, or new infrastructure – but rather by years and decades of educational efforts about animal welfare, health, and environmental impact. As awareness grew, so did consumer behaviour, leading millions of people to change their purchasing habits. It's a clear example of how demand *can* shape supply. Through our (online) shopping choices, we do have a degree of influence over what is produced and made available – while also recognizing the limitations of consumer-driven change.

Contagious potential and authenticity

'Just as ripples spread out when a single pebble is dropped into the water, the actions of individuals can have far-reaching effects.' This quote from the Dalai Lama may seem to contradict my earlier statements about the very limited impact of small individual climate actions. But consider it from another perspective: we should not use the very limited impact as an excuse not to reduce our own footprints. Because if our climate-friendly behaviour 'infects' and inspires others to follow suit, it could mark the beginning of *their* own climate journey – one that may ultimately lead them to a large climate handprint. After all, the US-American climate activist Bill McKibben, founder of 350.org and author of a dozen environmental books, probably also started his engagement by switching off the lights at home and reducing his individual carbon footprint.

This 'ripple effect' has been empirically proven. For example, scientists at the Potsdam Institute for Climate Impact Research were able to prove that PV systems have a contagion effect on the neighbourhood: As soon as a solar system is installed on the roof of a house in a residential area, the probability that others will be installed on other houses in the neighbourhood increases significantly.[182] This even works across district boundaries: for example, approval for the expansion of ground-mounted solar or wind power plants increases in one district if this is already being implemented in a neighbouring district.[183] There is also a proven contagion effect when it comes to flying, or more precisely, *not* flying: the more government politicians, celebrities and public figures refrain from flying, the more 'normal people' follow suit.[184]

All those who communicate and educate people professionally about the climate crisis, such as climate scientists, have a particularly important role model function.[185] It is extremely important and helpful for all of them to take steps to reduce their personal carbon footprint themselves in order not to lose the trust of their audience, their credibility and their impact.[186] Researchers found out, for example, after a climate lecture the audience will signal greater support for political climate protection measures if the person giving the lecture has a low carbon footprint themselves.[187]

From either-or to both-and

Some people may now be asking themselves the question: 'Should I reduce my carbon footprint or just focus on my handprint?' The answer is: It's not an either-or decision, but a both-and decision. But with the right balance of climate action with an immediate small effect and climate action with a delayed large effect. **Ultimately, it's about everyone finding their own personal mix of measures to reduce their footprint and increase their handprint.**

Nevertheless, I would like to offer a rough rule of thumb for a possible balance between the two approaches, based on the following train of thought: If we can realistically only reduce our carbon footprint by about a third, then we should also only invest a third of the time, attention and energy, that we want to devote to the climate issue as a whole, in reducing our footprint. The majority of our climate commitment – i.e. at least two thirds of our time, attention and energy – should, on the other hand, flow into increasing our handprint. In this way, we can exploit both the advantages of the footprint and the greater leverage effect of the handprint concept in an adequate cost-benefit ratio.

In any case, the effort, time and energy we put into reducing our footprint should be aimed at two things: firstly, to start with the biggest CO_2 levers in our lives – in other words, to tackle our most polluting activities as individuals. And secondly, we should try to achieve the greatest possible contagion effect with our footprint reduction, so that more and more people are inspired to start on their path to effective climate action and positively spill over to handprint enlargement.

How to do it better: with pareto-optimal 'big points'

Most people have a very busy everyday life that is overshadowed by other issues besides the climate crisis, such as the effects of the war in Ukraine or high energy bills and rent. As a result, our attention span for concerns about climate and environmental problems is limited despite the urgency of the situation.[188] (Incidentally, this phenomenon is known in psychology as the 'finite pool of attention'.[189]) This underlines the importance of the most effective approach to reducing our carbon footprints. After all, anyone who spends their limited time, attention, and precious 'willpower' on avoiding plastic bags, choosing eco-friendly online shopping, or turning off standby devices may focus on the crumbs while missing the larger pieces of the cake – not to mention the bigger handprint levers.

The German Federal Environment Agency therefore advises that it is better to stick to rules of thumb for the so-called 'big points' instead of constantly trying to get everything right in all areas.[190] These three 'big point areas' are mobility, food and housing.[191] So the practical thing is: we cover the biggest CO_2 areas in our lives with just three questions: 'How do I get from A to B?', 'What do I eat?' and 'How and where do I live?'.[192]

My suggestion is therefore a 'three-in-three rule': for the next three months, adopt a realistic number of *three new measures* from the ten suggestions in the following figure. This is then a footprint reduction according to the 'Pareto principle'. (The Pareto principle states that in many cases you can achieve the majority of the total possible benefit with just a small part of the total effort actually required). If we reduce our personal greenhouse gas emissions 'Pareto-optimally' and do not take the small eco-tips so excessively seriously, we can devote the rest of our attention, time and energy available for environmental protection to our handprints.

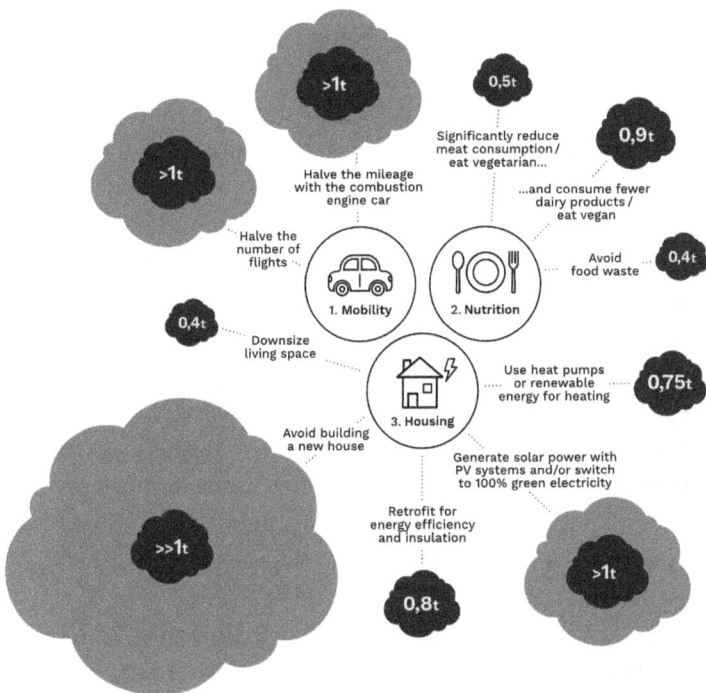

Figure 12: Three big point areas and the ten most effective measures to reduce one's carbon footprint (in tons of greenhouse gas emissions, largely based on the third part of the IPCC's AR6 report of 2022)

The illustration of our biggest sources of CO_2 in our private lives shows that we should worry much less about whether a light is still burning somewhere in the home, whether the dishwasher tab is wrapped in plastic or how many avocados we can still buy.[193] As we saw in chapter 3, the much more effective starting points for personal CO_2 savings are usually structural, major and long-term (purchasing) decisions – for example, moving house, building a new home or renovating, buying a car or a solar system, where we work or go on our next vacation or the fundamental decision on our diet. In this way, we can reduce our carbon footprint by one, two, perhaps even three tons or more with just three measures.

Looking at our fellow human beings and infecting them with footsteps

However, we should not just implement our three big-point measures quietly and secretly for ourselves over the next three months (and beyond), but should also focus on our fellow human beings so that as many people as possible can imitate us. The question should therefore not only be 'How many kilos can I get off my CO_2 scales?', but also 'How many of my fellow human beings can I motivate to take climate-friendly action?'. From this perspective, by taking your own steps to reduce your footprint, you are leaving footprints for other people to follow on their path to climate commitment. For this to work as well as possible, two components are needed: Firstly, our footprint actions need to be visible. And secondly, they should appear as attractive as possible (or at least not too costly).

Visibility is the easier part. Other people around us will automatically find out about many of our climate actions. For example, anyone who rejects meat dishes at lunch and opts for vegetarian or vegan alternatives, or who starts cycling to work every day instead of driving, will most likely be approached about it sooner or later. This was the case with a German friend of mine near Cologne, for example, when he had his house roof covered with solar panels immediately after the start of the Russian war against Ukraine. At first, his neighbors looked puzzled, but then more and more of them showed great interest and, after just a few conversations, also ordered solar panels. At the same time, however, he had also replaced the fossil gas heating system with a modern heat pump in the basement. When it comes to such less visible changes – for example, regional vacation trips, second-hand clothes or a new green electricity tariff – the simple rule is: do it and then *talk* about it. How else will other people find out about it and follow in your footsteps?

Presenting the CO_2 saving measures as attractively as possible, on the other hand, is more difficult. Even if we cannot immediately overcome all external, structural resistance, we can show by example: It can still be done! You can live on a small carbon footprint without having to give up all the comforts of life. What's more, you can even feel better and live a healthier and happier life. We can also highlight and emphasize the advantages and positive side effects of (reasonable) climate-friendly alternatives in conversations. For example, we can talk vividly about how we feel much fitter, have already lost weight and finally no longer feel back pain since we started commuting to work by bike every day. In general, when promoting climate-friendly actions in conversations, it is better to encourage and invite in a friendly and respectful manner, with charm and humor, rather than being evangelical, arrogant or aggressive. The Norwegian psychologist and climate communicator Per Espen Stoknes puts it this way: 'Any solution works much better if people want it, like it, love it, rather than implementing it out of a sense of duty or guilt, because of a regulation or out of fear of punishment.'[194] In simple terms: climate action should be as sexy as possible.

We can also make climate-friendly behaviour more attractive by reducing interpersonal obstacles for our fellow human beings. It is true that we can hardly avoid the do-gooder problem. But we should at least avoid the individual guilt, shame and hypocrisy debates, and put an end to them whenever and wherever they arise. For example, we can try to redirect the focus of the conversation to the handprint. More on how to successfully conduct such 'climate conversations' later in the next chapter.

Chapter 7: Enlarging the Handprint in Private Life

The first area of life in which each of us can increase our own climate handprint is our private life. We have two effective handprint levers at our disposal here: firstly, our influence on the people in our direct environment, including contacts or loose acquaintances beyond our family and close friendships. Secondly, how we invest money in banks, insurance companies or pension funds can influence whether a new solar park or a new oil pipeline is built somewhere in the world. However, before we can move these levers – and all the others in the following chapters – we need *time* as a basic prerequisite. So first I will address the important question: How many hours of our (free) time can and do we want to spend on climate action and enlarging our handprint?

The personal climate hour: time commitment for climate action

We all have 24 hours a day, 168 hours a week and around 720 hours a month. That sounds like a lot. But a large part of our time is taken up by our job, household chores, doctor's appointments, post, emails, text messages, possibly childcare and much more. As a result, most of us only have a very limited and precious amount of *really* free time. We want to and should use this time to do the things that bring us joy and give us a balance. And now climate action has been added to the mix? How is that supposed to work?

For most of us, taking action likely requires scheduling a regular, firmly anchored time slot in our calendars. This allows us to carve out mental space within our 'finite pool of attention' to focus explicitly on the climate problem, protecting it from other tempting distractions. This also has advantages: We don't waste this time with social media or short-term pleasures like the next TV series. Instead, we spend it on a deeply meaningful activity that is aimed at long-term well-being – both for us and for our fellow human beings and creatures.

In general, I suggest one 'climate hour' per week. This is because I assume that there are many people who can devote one hour a week (around 1 percent of our weekly waking time) to climate action. However, the exact amount of time will depend on your individual circumstances. For example, anyone who is confronted with an exceptionally large number of work or family commitments will rarely be able to devote a whole hour per week to the climate issue – in the worst case, none at all. The collection plate principle applies here: you give what you can and are willing to give. And you can also look at it this way: You usually get something in return. If I research a heat pump and then have it installed, I can save money. If I tell my neighbour about it, I've also had a nice conversation and strengthened the bond. And how good do I feel when I do something like this instead of refreshing my newsfeed ten times a day?

Climate education: inform yourself and prepare for action

Before you can act effectively, you need to plan – and before you can plan well, you need to be well informed. That's why I suggest spending the first few climate hours on climate education to find out more about the climate crisis, possible solutions and your own starting points. This applies to everyone, because you never stop learning about the climate crisis – no matter where you are on your personal climate journey. There are now countless educational offers on the topics of sustainability and climate protection – for example magazines, newsletters, podcasts, YouTube videos, documentaries, plays at theatres, art exhibitions, games, live lectures, discussion evenings and, last but not least, books (like this one). How and through which medium you prefer to learn is of course up to you. A selection of my personal highlights on climate education can be found in the appendix of this book.

Unfortunately, a growing number of people obtain their knowledge from dubious sources.[195] Since misinformation and disinformation are widespread, you should pay close attention to the reliability of information sources. Caution is always advised, especially with information shared online and through messenger apps, because false news has been proven to spread faster on the internet and reach many times more users than facts and truths.[196] A study by Brown University in the United States showed that around a quarter of all tweets about the climate crisis or global warming originate from so-called 'bots', i.e. small computer programs.[197] (This study was published in 2021; today, the numbers are probably even higher.) And the majority of these automatically generated posts are *against* climate action. Investigative research (for example by the NGO InfluenceMap) also indicates that fossil fuel companies are financing automated anti-climate action posts and comments like these via veiled detours.[198] This means: beware of online content without sources!

On the one hand, you can of course inform yourself on your own. For example, after my return from the UN Climate Change Conference 'COP25', I sat down in a public library almost every day from the beginning of January 2020 until the start of the coronavirus lockdowns to study the IPCC reports, summarize them in an easily understandable way and thus laid the foundation for my first climate podcast *Climaware* in Germany. But on the other hand, it is of course more fun if you go on a learning journey together with a climate buddy or group. And it also promises more learning success to exchange ideas, tips, experiences and best practices with similarly climate-conscious people and possibly receive practical support.

Breaking out of the climate silence spiral: talk more about the climate

Unfortunately, too little is still being said about the climate crisis during lunch breaks, dinners, barbecue afternoons, at parties or during random encounters. Logically, the impending end of the world as we know it is not exactly a party starter – more of a mood killer. And before you kill the mood or say the wrong thing, it's usually better not to say anything at all. This is true for most people, even if they are actually very interested in the topic. Most Americans, for example, don't talk or hear much about global warming, even though it is actually important to them personally, as surveys by the Yale Program on Climate Change Communication show.[199]

There is now even a term among experts for this phenomenon of preferring not to mention the 'C'-word in private encounters: the 'climate spiral of silence'.[200] The term goes back to the German communication scientist Elisabeth Noelle-Neumann, who introduced the concept of a 'spiral of silence' for controversial topics back in 1974. According to this concept, we humans tend to keep our opinions to ourselves if we assume that expressing them would catapult us into a minority position, social devaluation or even exclusion.[201]

And here comes the catch: the *actual* majority conditions in the social group or society are less important than the *subjectively perceived ones*. However, as vocal and self-confident fringe groups often dominate the debate on the climate issue and the 'silent majority' does not discuss it enough, the actual prevailing opinions are often obscured. As a result, most people significantly underestimate the willingness of their fellow citizens to protect the environment and climate.[202] And then they themselves demand and act less because they mistakenly believe that others (want to) demand and act even less. This is how the climate spiral of silence becomes a spiral of inaction.[203]

The most effective antidote to breaking the climate spiral of silence is quite simple: talk about the climate crisis as much as possible.[204] 'That's one of the most important things anyone can do,' says Anthony Leiserowitz, who heads the Climate Change Communication Program at Yale University.[205] And the climate researcher Stefan Rahmstorf told me that talking more about the climate crisis is his favorite tip when people ask him what they can do about it.[206] Then many people may not only realize that their neighbour is also very worried about the climate crisis, that their uncle would be prepared in principle to take meaningful climate action measures or that their colleague is already doing many climate-friendly things. Under the right circumstances, in-depth climate talks can be the beginning of a more intensive engagement with the climate crisis for some people, a greater motivation to act and ultimately a larger handprint.

Use and create opportunities for climate conversations

Climate conversations can be triggered in two ways: either you raise the topic yourself in a suitable situation or you increase the likelihood that others will raise the topic. There are many opportunities and thematic hooks in everyday life: For example, it could be the current heatwave, a news headline, a much-discussed talk show debate, something you have newly learned yourself or a meeting with an inspiring person, a regional climate action campaign or your neighbour's family's new heat pump. However, opening a climate conversation yourself requires not only a thematic hook, but also a sure instinct and the right timing. The people taking part in the conversation should have time and, in principle, be ready for an in-depth discussion. On the other hand, anyone who has acute worries, is emotionally upset or is currently hungry will certainly not be very willing to talk about the climate issue.

To increase the likelihood of being approached by others, it is enough to make a visible eco-friendly contribution yourself – even if this only involves reducing your own carbon footprint. The British organization Climate Outreach found through surveys that most conversations about the climate issue start with the fact that at least one person involved in the conversation has done or is doing something to protect the environment or climate.[207] So, 'do good and let people know about it'.

Further increasing the impact of climate conversations

The two ways of triggering conversations about the climate crisis described so far are about spontaneous conversations. However, climate conversations can be even more effective if they are specifically prepared, i.e. if they are planned. For example, you can start with the following two questions: firstly, 'When and where do many people I know come together in their free time?' and secondly, 'Who do I know who is more influential than me?'. If you ask yourself these two questions and think about how, when and where you can motivate these people to take climate action, you will not only talk more about the climate, but also in a more targeted and effective way.

With regard to the first question, I would like to briefly touch on the digital world: Most of us can reach hundreds to thousands of people with just a few clicks, particularly via social media and messenger services. Talking to friends or acquaintances online about the climate crisis and presenting our own climate action, for example in our profile or with stories, is therefore an opportunity to exert influence that we should not forget. But of course, relatives, friends and acquaintances also come together in larger groups offline on certain occasions. These are usually celebrations such as weddings, birthdays or traditional festivities like Christmas. If you show particular sensitivity and take the festive mood into account, these occasions can be used to address the climate crisis.

A personal example: To celebrate my 30th birthday, I invited all my friends to a 'climate carnival birthday party' in Cologne – a contradictory motto at first glance because, as I said, the word 'climate' is not really a big party starter. But it worked anyway – and surprisingly well. On the one hand, because the program was split into two parts: first there were two short climate lectures in the afternoon, moderated discussions and plenty of time for open exchange in a relaxed setting and then the party in the evening. Secondly, because the communication was deliberately positive and solution-oriented and focused on the question 'What can I do?'. By emphasizing the climate handprint, among other things, we were able to avoid the usual debates about renunciation, guilt, shame and hypocrisy. It was an experiment that taught me something: With a little courage, preparation, empathy and tolerance, you can also integrate the topic at celebrations.

The second question for even more impact in climate conversations ('Who do I know who is more powerful and influential than me?') is initially about my private environment – for example, a cousin who works in the Ministry of Finance, a neighbour who is a professor at a university, a friend who holds a management position in a company, or a team member in a sports club who is involved in local politics or knows the mayor. If you manage to engage such influential private contacts in climate conversations that touch them emotionally and motivate them to do even more for climate action, you may be able to significantly increase your climate handprint.

Not just talking more, but also talking better about the climate

Up to this point, I have pretended that we live in an ideal world in which everyone communicates in a respectful, fact-based and friendly manner. However, most people have probably already experienced first-hand that conversations about politics, and in particular about an emotionally significant topic such as the climate crisis, can lead to ambivalent results. Ideally, everyone involved will leave the conversation inspired, more knowledgeable and more motivated. However, it is not uncommon for climate talks to be very frustrating or even emotionally hurtful.

There are many reasons why climate talks in general are a tricky business: The climate crisis is incredibly complex and many of the solutions will create conflicting goals and both winners and losers. (This is why the green transition must be a just transition and involve some kind of redistribution). As mentioned, the exaggerated individualization of the climate crisis is also leading to widespread debates about guilt, shame, renunciation and hypocrisy. However, one of the main causes is that many people who see their identity, values, freedom, beliefs or positive self-image threatened by climate protection measures (unconsciously) ramp up their psychological defense mechanisms. For example, what is known in psychology as 'motivated reasoning': they then think or argue in a strongly biased and targeted way so as not to have to deal with the emotional or political consequences of the radical need for climate protection.[208] Under such conditions, a calm, rational and objective conversation is obviously difficult or even impossible.

We have already seen this type of self-defense, or emotional adherence to the status quo, in the case of the Still Unwilling in Chapter 2: some doubt the effectiveness of climate protection measures with statements by pseudo-experts, logical fallacies, cherry-picking or conspiracy myths.[209] Others use the argumentation tricks of the common *discourses of climate delay*: These include shifting any responsibility for action away from themselves (for example, exclusively to the rich, large corporations or China), pushing for non-transformative pseudo-solutions (for example, large-scale Carbon Capture and Storage, green hydrogen, synthetic car fuels or geoengineering), the overemphasis on the disadvantages of climate protection ('... is not a perfect solution', '... damages the economy', '... burdens the poor' and the like) and, more recently, the climate doomism already described ('It's too late', '... hopeless', '... impossible' and so on).[210]

The tricky thing about these argumentation patterns used by climate action delayers and blockers is that most of them contain a grain of truth. However, they typically only shed light on the issue in a one-sided and incomplete way or neglect existing solutions or the consequences of delay and inaction.

To avoid all these potential conflicts, we should not just talk more about the climate issue, we should talk better.

Climate communication: Ten tips for better climate conversations

If you keep a few basic rules in mind and truly internalize them, you can already make significant progress toward more fruitful climate conversations. I originally compiled the following ten tips for the podcast *Über Klima Sprechen* (Talking about climate), together with climate journalist Christopher Schrader and the team from the German NGO Klimafakten, based on their handbook *Über Klima Sprechen*, the Talking Climate Handbook by the British organization Climate Outreach, my own experience and various expert discussions:

- The basis of every climate conversation is **respect.** This means taking time for the conversation, approaching the other person with a calm, open, interested and friendly attitude, and letting them finish. It is essential to avoid coming across as bossy, judgmental or aggressive. This also applies to body language of course.

- At the beginning, you should limit yourself to **questions** instead of immediately presenting your own views and arguments and lapsing into a monologue. The questions should be as open and personal as possible in order to find out where the other person is at, where they are coming from, what moves, worries and motivates them. And then it is important to listen carefully. The thought 'I am here to learn, not to teach' helps me personally.

- If you get a feel for the other person's reality and values, it's easier to find a **common denominator** and focus on what connects us rather than what divides us: What values are important to both of us? What concerns do we share? This common ground then also provides a helpful point of return should the conversation take a turn for the worse.

- If possible, the climate conversation should not drift into overly abstract concepts or complicated technical talk, distant places or too far into the future. In order to avoid psychological distance and instead create emotional closeness, it is better to focus on the **here and now** using understandable language.

- At a certain point in the conversation, it can be helpful to open up and **tell your own story** authentically and honestly: When and how did you yourself become aware of the environmental problems? How do you feel when you consciously think about the climate crisis? Where do you still have difficulties, for example in behaving in a sustainable way?

- We should generally focus on **positive and motivating narratives**: for example, a desirable vision of the future, climate solutions, role models, positive examples, options for action and what we are already doing ourselves. This reduces paralyzing feelings of fear and powerlessness and instead awakens feelings of self-efficacy. Of course, this does not mean romanticizing the seriousness of the situation through unrealistic wishful thinking.

- When discussing personal options for action, we should not fall into the trap of individualized debates about renunciation, guilt, shame and hypocrisy and talk too much about the little eco-tips of everyday life. It is better to talk about ways to increase your personal **climate handprint**.

- If, despite the above tips, the conversation does become emotional and **feelings** such as fear, sadness or anger well up, it helps to address them gently and empathetically, give them space and reflect on them together. In extreme cases, however, it may be advisable to change the subject and calmly address a particularly strong emotion again at a later time.

- During the conversation, the other person may fall into the typical argumentation patterns of the **discourses of climate delay** mentioned above. Simply countering them with facts usually doesn't help. It is much better to limit yourself to asking questions: Why are you worried about this? What alternative do you propose? How realistic do you think this is? In this way, you can find out whether their concerns are well-founded and justified or just 'motivated reasoning' with smokescreens and fig-leaf arguments.

- With the latter, the question arises: **stubborn or just confused**? With stubborn people, patience and kindness are required. In any case, it is important to find a positive ending to the conversation, even if you don't agree. And if that is not possible, you can of course break off the conversation and leave at any time out of self-care or even self-protection. After all, it takes too much time and energy to try to have constructive conversations with the small minority of particularly stubborn or even hateful delayers or deniers. (Of course, this also applies to hateful or anti-democratic climate action advocates). It is better to focus on those who are interested and open – or merely confused.

The family: a special case

These tips are particularly useful when you are just starting out with climate conversations. At some point, you realize that the conversation patterns become ingrained and repetitive. After a few conversations, you will know exactly how your loved ones feel about climate protection, what worries and concerns they have, what experiences, arguments or stories they typically cite in response to certain questions and where their emotional trigger points are.

This is especially true for close relatives. The family is a special case because most people are firmly bound to this community of fate. There are two general options here: Either there is fundamental agreement on the climate crisis and the necessary climate action measures. This makes the family a place of refuge that provides strength and comfort in the midst of the emotionally stressful planetary emergency. Or the family members evaluate the urgency of climate action and the necessary measures too differently, so that the climate issue becomes a recurring point of conflict or contention. This makes the family more of an emotional burden in one's climate-conscious life.

In the first, optimistic case, conversations about the climate crisis can possibly develop a collective motivation to act and a self-effective sense of unity. It may even be possible to change your entire life to be as climate-friendly as possible, like the Pinzler-Wessel family in chapter 6. Research shows: The parent-child relationship plays a crucial role here, as children – and daughters in particular – can have a strong influence on their parents' climate awareness.[211] However, it is of course not only children who influence their parents in terms of sustainability, but also the other way around – above all through environmentally friendly role models and climate-conscious parenting. One could say, as a parent, you automatically increase your climate handprint by exemplifying and teaching your children climate-friendly values.

Unfortunately, there is also the second, unpleasant case of a recurring conflict or argument – even though (or precisely because) the family members are so close to your heart. In this case, you should pay particular attention to the first three of the previous ten conversation tips: Respect and tolerance, questions instead of statements and the common denominator of unifying values. If this doesn't help either, and heated, emotional or hurtful debates continue to arise, it's better to nip the topic in the bud next time. For example: 'Please let's not talk about this again. We already know where we stand and how we feel about it. Let's talk about ...' It's okay to let stubborn relatives stand by their rigid point of view for the time being, graciously and patiently, because trench warfare within the family costs many times more time, nerves and energy than we can afford in view of the already stressful planetary emergency. Choose your battles! After a few months, you can make another tentative attempt.

Impact investing: what does my money do in the world?

Science journalist Christopher Schrader wrote in the epilogue of his handbook *Über Klima Sprechen* on climate communication: 'You can't do it without talking, but you can't do it with *just* talking either.'[212] He is right, of course. So now we come to the second central handprint lever in this chapter. This requires a little abstract imagination at first, but should then seem a lot more tangible and concrete than the previous topic of climate communication.

By abstract imagination, I mean being roughly aware of the chains of effects in the global financial system. Most people know roughly how much money they have in their bank account. But probably only very few know what exactly this money does in the world. Yet the figures on your personal bank statement have a very real impact somewhere in the world. For example, personal money can help finance the expansion of an open-cast coal mine in Australia, a new oil pipeline in Canada or a new solar power plant in Spain. So we should all be aware of this as a matter of urgency: Money rules the world – often silently and hidden, even if you don't feel it directly in everyday life.

We let them invest our money: banks, insurance companies and pension funds

The big corporations that profit from the burning of fossil fuels have already been given their comeuppance in this book. However, their disastrous business practices are only possible on the current scale because banks support them with loans, among other things. The report *Banking on Climate Chaos 2024* reveals that the world's 60 largest banks provided 705 billion US dollars to companies engaged in the fossil fuel industry in 2023 – an amount exceeding the Gross Domestic Product (GDP) of Poland.[213] Since 2016, the year the Paris Climate Agreement came into force, these banks have collectively funneled about 3.3 trillion (that's three thousand three hundred billion) US dollars into companies with fossil fuel expansion plans. UN Secretary-General Antonio Guterres has strong words for such business practices: 'Investing in new fossil fuels infrastructure is moral and economic madness.'[214]

Among these 'banks for climate chaos' are well-known names such as HSBC and Barclays (both headquartered in London) or JPMorgan Chase, Citigroup, Bank of America, and Goldman Sachs.[215] You should do some research and carefully consider whether you want to be a customer of banks that finance climate chaos. But even if your own bank is not on a list of the most climate-damaging financial institutions, it is worth asking whether and to what extent it still invests in projects or companies with activities in fossil fuels. Perhaps, with a little pressure and together with other customers, you can even get the internal investment guidelines changed, as happened at HSBC at the turn of 2023, for example: Europe's largest bank announced that it would no longer finance new oil and gas fields as part of its updated climate strategy.[216]

However, you can send a stronger signal and drive change more quickly by voting with your feet and switching to a climate-friendly bank as a customer. The consumer organisation Which? ranked the three 'greenest' banks in the UK as The Co-operative Bank, Nationwide Building Society, and Triodos Bank. Triodos, an online bank, was rated the best overall. It only lends to organisations that create a positive impact for people and the planet, supporting renewable energy, sustainable farming, and social housing, while rejecting investments in fossil fuel industries.[217]

Where our pension money is invested is important, too. UK pension funds have investments totalling three trillion pounds. Unfortunately, some 88 billion of this is in fossil fuel companies, meaning that an average of £3,000 per pension holder is invested in the likes of BP and Shell, according to a recent report by the group Make My Money Matter.[218] You can check with your pension fund where your investments are held. If they contain fossil fuel companies, you can contact your employer or fund directly and ask them to switch to a sustainable investment strategy. Your climate handprint will increase considerably if the demand is implemented. If they do not change track, you could look into transferring your pension to a new, more environmental provider. Make My Money Matter rates Aviva, Legal & General and Nest as the most climate-friendly pension providers. And of course: always seek professional financial advice before transferring a pension or making investment decisions.

We invest ourselves: shares, funds, impact investing, etc.
Many British people invest their savings directly, and 21 percent of assets under management in the UK are held by retail (i.e., ordinary) investors.[219] Many Britons, and citizens around the world, can therefore directly decide for themselves

where their money goes in the capital market – whether it finances more greenhouse gas emissions or contributions to climate solutions.

However, not everything that says green is actually green: Keep your eyes open when investing or buying shares. Unfortunately, greenwashing is widespread among the rapidly growing number of ESG and climate funds.[220] This can be seen in a corresponding study by InfluenceMap.[221] According to this non-profit transparency organisation, over 70 percent of the almost 600 ESG-labeled funds examined scored negatively when it came to investment alignment with the Paris climate goals. Even among the financial products explicitly labeled as 'climate funds', only around 45 percent received a positive rating in this regard. UN chief Antonio Guterres comes to mind again, this time with the words 'Some government and business leaders are saying one thing, but doing another. Simply put: they are lying.'[222]

However, such figures should not discourage us. With the help of analyses by independent NGOs and a little time and diligence (remember the climate hour), you can fight your way through the jungle of 'green' ESG and climate funds and in some cases identify real climate champions that not only promise a good return, but also make an effective contribution to achieving the Paris climate targets. Because the positive message of InfluenceMap's report *Climate Funds: Are They Paris Aligned?* is: Yes, some of the climate funds examined received enormously positive climate ratings.[223] To summarize: Some are lying their way to green, but not all. And we can identify these true climate pioneers and support them and our handprint with money.

This is especially true for many start-ups and young companies that want to bring new technologies or solutions for the climate transition to the market and spread them. For example, a good friend of mine founded Orbio, a startup providing data on large methane leaks worldwide, and needed long-term financing after the initial cash injection from a funding programme. Such venture capital usually comes from professional investors, so-called 'business angels', or venture capital (VC) funds, private equity funds that specialize in startups investments.

However, if you have basic investment knowledge and are aware of the relatively high risks, small investors can now also provide promising climate start-ups or energy transition projects with urgently needed financing through crowdfunding.[224] In this way, you join forces with many other small investors so that you can invest together in one or more start-ups (for example, automatically via a crowdfunding platform). The philosophy behind this type of impact investing is that by investing out of conviction, you make a measurable contribution to solving social or ecological problems while still generating a return.

And one last point on the subject of shares: as a shareholder, you usually automatically receive voting rights for important decisions at the shareholders' meetings. You can also usually propose your own resolutions, raise sensitive issues or ask questions in public. As a shareholder, you can use all of these opportunities to push for climate action and increase your handprint. In 2021, for example, a study by the auditing firm Ernst & Young showed that the chances of success are increasing: at the one hundred largest US companies, the number of successful sustainability proposals from shareholders has increased more than six-fold since 2016.[225] This shows that if the green transformation of the US economy is not currently being propelled top-down by the Trump administration, it can instead be driven bottom-up by citizens and the private sector.

Chapter 8: Enlarging the Professional Handprint

Fridtjof Detzner founded his first company at the age of just 16, which later became the successful website builder Jimdo.[226] In the almost two decades that followed as managing director, Fridtjof measured the success of his company mainly using conventional business figures, primarily based on turnover and profit.

But then he traveled through Asia with a film crew from Deutsche Welle (DW), Germany's international state broadcaster. Their aim was to record a documentary series about founders whose companies make a positive contribution to the 17 Sustainable Development Goals (SDGs) of the United Nations. Fritjof's travels took him to a region in India that some call the 'Suicide Belt' due to the statistical correlation between disrupted weather patterns and high suicide rates. (A widely cited study from 2017 attributes over 50,000 suicides in India to climate change).[227] There he met an Indian farmer who, after three consecutive failed harvests, had desperately tried to kill himself by taking pesticides so that his son would at least receive a small orphan's pension. He survived – and now changed Fridtjof's life forever with his story.

Back in Germany, 'Fridel' first had to process his many impressions and experiences – especially his encounters with all the suffering and misery. But then it was clear to him: from now on, he would only invest his time and resources in companies that were doing better than today's economy. This gave rise to his new motivational phrase, which has been emblazoned in large letters on the wall in his office ever since: 'Build something the planet needs'. Today, Fridel uses his knowledge and experience, together with his co-founders of the Planet A fund and through their investments, to finance start-ups that have a measurably positive impact on the planet and people.

To what purpose do I dedicate my working time and energy?

After Fridel told me the story of his personal transformation, he posed another question during our podcast interview that has stayed with me ever since: 'Who or what am I actually dedicating my labour to?'[228] It's an important question, because we spend an average of 80,000 hours of our precious lifetime on (paid) work. And during this time, we can use our talents, skills, knowledge and experience to achieve very different things in the world – even if we may not directly feel the effects of our professional activities ourselves.

As an engineering graduate, I could have either worked for an oil company planning more efficient fracking wells or for a renewable energy project developer planning larger wind turbines. The day-to-day work would probably have differed only slightly: working with planning software, Excel spreadsheets and emails, plus meetings, phone calls and site visits, interspersed with lunch breaks and team events with the other engineers. However, the effects on the further course of global warming could hardly be more different.

The question of how to use your working time wisely therefore provides further food for thought: What does my job do in the world? What factors fulfill me professionally? What motivates me – apart from salary and possibly status – to go to work every day? What is my vocation and my deeper purpose in life? Why and for what do I get up in the morning?

Find your own climate ikigai

Those who find authentic answers to these profound questions and align their lives accordingly can reach a state of greater life satisfaction and fulfillment. This state is called 'Ikigai' in Japan. The British entrepreneur, coach and blogger

Marc Winn combined the Japanese Ikigai philosophy with a Venn diagram for more meaning in life. This concept asks four questions: 'What do I love?', 'What am I good at?', 'What can I get paid for?' and 'What does the world need?'. At the intersection of the four questions lies the area where you can find or fulfill your professional purpose in life.

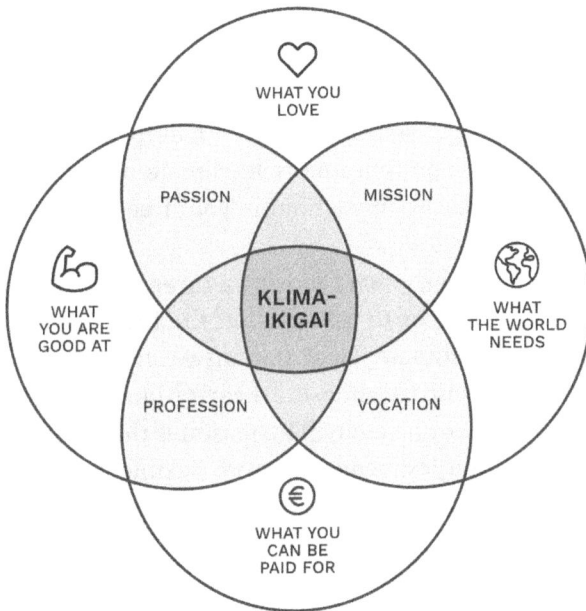

Illustration 13: Climate ikigai diagram (freely adapted from Marc Winn)

Normally, most people's jobs lie at the intersection between the two questions: "What am I good at?" and "What can I get paid for?". Some strive for a little more joy or passion in their job and add the question of what they really enjoy or love. However, I find it very powerful – and extremely appropriate in view of the planetary climate emergency – to use the fourth question to view the job a little more as a mission and calling to preserve our livelihoods. The German word 'Berufung' (vocation) comes from 'Rufen' (calling) – and our planet is urgently calling us for help with droughts, forest fires, storm surges and melting ice. Anyone who finds an activity that lights a fire in their heart and which also contributes to stopping the wildfires in the hearts of our forests can thus find their own personal climate ikigai. (Of course, this is only achievable if you are not trapped in precarious paid work and have some level of choice in your profession).

This not only allows you to achieve a meaningful, satisfied and fulfilled basic state in life. But with a 40-hour working week and the use of your professional handprint levers, you contribute at least 40 times more to climate protection than with just one weekly climate hour in your free time.

Keep your eyes open when choosing a career

One of the best times to ask yourself the four questions of the professional climate ikigai and answer them honestly is probably when you are choosing a career immediately after school, training or university. The reason is that at this stage you are still young, so generally more flexible, independent and idealistic, and have lower costs to cover than later in life. In addition, you still have the entire 80,000 working hours of your career ahead of you. This is probably one of the reasons why United Nations Secretary-General Antonio Guterres gives speeches to graduating university classes. His message for the class of 2022 was simple: 'Don't work for climate wreckers. Use your talents to drive us towards a renewable future.'[229] A growing number of well-educated young people are heeding his call and want to make as big a positive environmental impact as possible.[230]

A friend of mine, Janna Hoppmann, is a good example of this: during a trip to Brazil on a container ship, Janna met Pablo, a sailor from the Philippines. His reports of flooded houses and other catastrophic climate consequences in his home country touched her emotionally so much that she developed a vision: leaders in Europe making courageous decisions for global climate justice in order to secure the basis for a good life for people like Pablo worldwide. And she decided to start a new profession straight after graduating in psychology: As a climate psychologist and founder of the social business ClimateMind, Janna now trains leaders from business, politics, public administration, and NGOs across Europe and beyond in transformational skills such as authenticity, compassion, activating climate communication, courage, and resilience. She designs innovative dialogue formats that bring decision-makers and frontline communities together to co-create solutions for the climate and biodiversity crisis. In just a few years, she and her team have already enhanced the impact of over 5,000 decision-makers.[231]

Like me, Janna followed an academic path to find her climate ikigai. But when it comes to choosing a career, it can also be a manual climate job. For the key energy transition to succeed, we need significantly more people in the skilled trades who can refurbish buildings and install all the necessary solar systems, wind turbines, heat pumps and charging stations. In the UK alone, hundreds of thousands of skilled workers are missing for the green energy transition forward.[232] In addition, increasing climate disasters will mean that many more people will be needed in the future for rescue services, fire departments, disaster control and healthcare – even if we manage to stop global warming well below 2°C in this century.

Switch your job to the green side of the Force

However, the UN chief's words 'Don't work for climate wreckers' are now not only being taken to heart by career starters, but also by people who are in the midst of life and in some cases had flourishing career prospects – such as Caroline Dennett. She ended her eleven-year working relationship with Shell in May 2022 because the oil company was continuing to expand its oil and gas production, contrary to scientifically calculated climate target paths.[233] 'I can no longer work for a company that ignores all the alarms and dismisses the risks of climate change and ecological collapse,' she wrote in her resignation email, which she sent to the entire management team and around 1,400 Shell employees.[234]

Such a drastic step and the subsequent professional reorientation naturally require a good deal of courage. But this courage usually pays off, at least personally. I suspect her resignation was a real relief for Caroline Dennett after years of doubt and remorse – perhaps even a form of salvation.

But what if the situation is not as clear-cut as it was for Caroline Dennett? What about all the jobs in the grey area that are neither extraordinarily harmful to the climate nor do they promote climate protection particularly strongly? Even then, of course, it is possible to take a career change in order to do work that clearly accelerates climate protection. Fridtjof Detzner's transformation at the beginning of this chapter proves this, as does the example of another friend of mine, Daniel Obst.

Daniel had already been working as a project manager and digital transformation executive at various companies for more than 15 years when he suddenly questioned his entire professional life and work. As a father of two, the Fridays For Future climate protests triggered a process of awareness about the planetary emergency, and he realized: 'My children will later ask me what I did about the climate crisis. Then I want to have good answers for them. That's why I have to act *now*, for their future.' So one night in January 2021, he simply got started with a kind of climate hour: what began as a blog for climate action in his spare time took on more and more professional traits as the reach of his online contributions grew – until he was finally even recognized by LinkedIn as a 'Top Voice' for sustainability. Just one year later, Daniel quit his well-paid job at a large German insurance company and quickly found his new vocation as a management consultant at a sustainability agency. Since then, he has been supporting companies in their sustainability transformation and giving talks and workshops on how to achieve the 17 UN Sustainable Development Goals.[235]

Another example of a late career change from the 'grey area' to the 'green area' is Janine Steeger.[236] She managed to realize her childhood dream: For many years, she hosted the TV show *RTL Explosiv*. But then the 2011 Fukushima nuclear disaster happened during her pregnancy, and Janine began to focus more intensively on the issues of sustainability, environmental and climate protection. As a result, her concern for the state of the planet grew, as did her motivation to act – so much so that she not only made her private life more eco- and climate-friendly, but also approached her RTL superiors with program concepts and suggestions on sustainability issues. But in vain: she was met with nothing but rejection. (Obviously Janine was ahead of her time, because the programme *Klima Update* has been running regularly on RTL since 2021).[237] As Janine was unable to implement any changes at RTL, she left her former dream job in 2015 after more than 15 years in television to become a self-employed sustainability presenter and speaker. She talks about her personal transformation from 'eco-sinner Janine' to 'Green Janine' in her book *Going Green – Why you don't have to be perfect to protect the climate*.[238]

In addition, there are of course many, many other professional handprint role models who have taken climate change as an opportunity for a career change. Today, they are all paid to work part-time or full-time for climate action and a future worth living – and the number is growing.[239] At the same time, portals for 'green' jobs are literally springing up – for example GoodJobs, greenjobs, Jobverde, Climatebase, inClimate, Terra.do, climatEU, Environmentjob or Work on

Climate. If, like a third of all European employees, you are already thinking about leaving your current job in the near future, you can look for a 'green' job on these portals, among others.[240]

Reduce working hours

I am aware that not every employee can or wants to consider a job change just like that. Some life situations or special circumstances simply don't allow it. However, if you can't or don't want to switch to a true climate action job, there are still two ways to make a positive contribution to protecting our environmental life support systems through your job: firstly, you can reduce your working hours – if that's possible.[241] And secondly, you can try to bring about climate-friendly changes *within* the company you work for.

At first glance, the first proposal probably seems neither plausible nor particularly attractive: why should it be beneficial for the environment and our climate if you only work part-time and earn less money? The answer has two parts, one about *less* and one about *more*. With regard to the less, a reduction in working hours offers the opportunity to get off the consumerism hamster wheel that has been spinning ever faster during the past decades. For many, it has led to overconsumption, stress and rising CO_2 emissions rather than greater satisfaction.[242] We also know from Chapter 6 that the level of income usually goes hand in hand with the size of the personal carbon footprint. People who only work 20 or 30 hours a week and thus voluntarily forego part of their salary (without having to worry about their livelihood or leave basic needs unmet) are therefore very likely to reduce their per capita emissions – and feel less stress.

And that brings us to the *more*. The positive side effect is that shorter working weeks lead to greater well-being and improved health. This has been proven by numerous large-scale studies, for example recently in the UK and Iceland.[243]

Those who work less can therefore not only live more frugally and calmly, but also become happier. In addition to more time for the things and people you love, this also means more time for private or voluntary social or environmental commitment.

In addition to a reduction in weekly working hours, there is another option to free up time for climate commitment. Many companies now offer their employees the option of taking a sabbatical – an unpaid break from work that typically lasts several months to a year. For example, service designer Ella Lagé from Berlin took such a climate sabbatical in 2016. On the one hand, she wanted to become even more involved in the Fossil Free Berlin campaign and, on the other, to persuade her pension fund, Versorgungswerk der Presse, to stop investing in fossil fuels.[244] Thanks in part to Ella Lagé's time and tireless commitment, a few months later the state of Berlin decided to withdraw public funds from companies in the fossil fuel industry and Presse-Versorgung withdrew its capital from coal financing.[245]

Climate action 'top down' in companies

For all those who cannot or do not want to change jobs or reduce their working hours, there is still the option of making climate-friendly changes to their current job. After all, companies can and must be part of the climate solution. This also applies to most companies with previously carbon-intensive business models. According to a 2020 study, the potential of so-called 'non-state actors' – such as cities and companies – is remarkable: if all companies apart from cities and municipalities implement their climate targets, up to 5.5 percent (i.e. 2 billion tons) of global greenhouse gas emissions can be avoided by 2030 in addition to national climate protection plans.[246] At the same time, according to the World Economic Forum (WEF), 'climate leaders can attract and retain better talent, realize higher growth, save costs, avoid regulatory risk, access cheaper capital and create new sources of value for customers.'[247]

The management, supervisory board and owners of companies have a particular responsibility in this regard, as they ultimately decide 'from above' where things are going. Some of them interpret their responsibility in purely conventional economic terms, focusing on quarterly figures and relatively short-term profits. Other entrepreneurs, however, feel a great responsibility for future generations, like Fridel from the beginning of this chapter.

Mission climate neutrality – also for suppliers and customers

Many entrepreneurs are now thinking about how they can become true climate leaders and accelerate society's climate transition. However, they should not fall into one of the widespread carbon footprint traps for companies. I would like to briefly introduce five of these traps.

Trap 1

Climate action in companies is not done with a little more waste separation, double-sided paper printing, LED lights, facade greening or the note 'Think before printing' in the email signature. If you don't just want to scratch the surface, you will soon realize that sustainability is a far more complex issue. It is therefore essential that companies either build up the necessary expertise internally – for example with trained climate and sustainability managers – or bring in external expertise – for example from a reputable climate protection and sustainability consultancy. After all, for almost all companies, it is more about a far-reaching transformation of business processes than just more insect hotels, a few plants and trees.[248]

Trap 2

When measuring and reducing greenhouse gases, you should adhere to the international standard: the 'Greenhouse Gas Protocol'.[249] This means not only focusing on CO_2 and the company's *internal* emissions (referred to as 'Scope 1 & 2' in technical jargon), but also measuring, publishing and reducing all greenhouse gas emissions (i.e. also 'Scope 3').[250] This holistic greenhouse gas balance also includes emissions from energy consumption, waste and transportation, the provision of raw materials, investments, business travel and product use by customers. This is because a company's responsibility and influence do not end at the company gate, but extend along the entire supply chain – 'upwards' to all suppliers and 'downwards' to all consumers.[251]

Trap 3

Entrepreneurs should not just view climate protection short-sighted and as an isolated issue. Instead, they should fundamentally question their business model without indulging in excessive technological wishful thinking or pseudo-solutions, and think: What exactly do our products or services achieve in the world? What can a truly sustainable vision for my company look like without using any fossil fuels? And how can we get there as soon as possible? Then we need a science-based climate strategy, including ambitious climate targets and concrete CO_2 reduction steps, for example with the help of the Science-Based Targets initiative (SBTi). [252]And finally, this climate strategy should be embedded in a comprehensive *sustainability* strategy that takes into account all 17 UN Sustainable Development Goals. The ultimate goal of every company must be sustainability and climate neutrality in all aspects and across the entire product life cycle – and preferably well before 2050.

Trap 4

CO_2 certificates for reforestation projects or similar are not a panacea, but merely a last resort. This means that they should only be purchased in addition to current reduction measures in order to compensate for previously unavoidable emissions. As an alternative, a company can of course also start its own offsetting project – for example for nature restoration and reforestation or technical CO_2 reduction. In any case, the utmost care must be taken when selecting a reputable provider and implementing the offsetting projects. Research by *The Guardian* and German *ZEIT* newspaper recently showed that over 90 percent of the rainforest CO_2 certificates issued by one of the world's leading providers in this field are more or less worthless.[253] Anyone who calls themselves 'climate neutral' on the basis of such certificates is engaging in greenwashing.[254]

Trap 5

As with private individuals, probably the most widespread footprint trap is for companies to focus solely on their own carbon *footprint*. However, companies can substantially increase their climate *handprint* as well. Broadly speaking, this can be done in three ways: by developing new markets, through social commitment and political influence.

Enlarging the company's handprint

The first of these three handprint areas is closest to the conventional understanding of businesses. A company can produce or provide new products or services in order to displace climate-damaging products and services on the market and thus accelerate the climate transition for society as a whole. For example, the Swedish steel manufacturer SSAB was able to present the world's first 'green' steel in 2021. And the logistics company Maersk operated the world's first cargo ship with climate-neutral methanol in 2023.[255] Both of these are hugely important innovation milestones for steel production and international shipping, which have been stubbornly emissions-intensive up to now. A traditional automotive supplier, for example, could also set up battery production for large-scale storage systems. Or a startup that helps users fill out and submit their tax returns could offer software that makes it child's play to deal with the bureaucracy that comes with installing a new solar system.

But it doesn't always have to be new technology, because new *old* stuff can also increase the company's handprint. For example, IKEA breathes a second life into used furniture in its 'Circular Hubs' as it moves towards a circular economy.[256] And the Too Good To Go app, which allows you to pick up leftover food at a low price from supermarkets, restaurants, bakeries and the like, prevents food waste.[257] Companies can also forge partnerships and alliances with their competitors in order to set binding social, environmental and climate standards for the largest players in an emerging economic sector. One of many examples of this is the Global Battery Alliance for the socially responsible and resource-conserving production of car batteries.[258]

The second area of corporate handprint influence is their social impact, including through social commitment. Some of the financial or other donations that have so far usually gone to local social institutions such as football clubs, kindergartens, retirement homes or homeless charities could also be donated to climate projects or environmental organizations. For example, a supermarket chain could donate cargo bikes to disadvantaged families, a bank could hand over a cheque to a local climate action initiative, a large law firm could offer free legal advice to climate NGOs, and so on.[259]

The ideal social solution would be to transfer the company to a (environmental protection) foundation, as Patagonia founder Yvon Chouinard did in 2022.[260] Or at least to ensure the company's profits benefit many people and the planet, not just a few owners. For example, the Google competitor Ecosia channels 80 percent of its profits into reforestation and the remaining 20 percent into renewable energies, regenerative agriculture and grassroots movements – instead of cashing out high dividends to stock owners.[261] In this way, the online search engine has already planted over 230 million trees and created a huge climate handprint.[262]

But beyond monetary or material aspects, companies also have an influence on society that should not be neglected – more precisely, on the knowledge, opinions and attitudes of the workforce and their families. In addition to traditional training and development courses, more and more companies are now offering their employees further training opportunities in sustainability and climate education during working hours. I already assisted a few companies to set up a tailored sustainability learning journey with monthly (online) lectures for the entire workforce.

Finally, the third area of corporate influence is politics, where company owners and executive leaders can either delay and dilute or accelerate and strengthen the societal climate transition through lobbying. The British NGO InfluenceMap and the We Mean Business Coalition bring multinational 'Corporate Climate Policy Engagement Leaders' (like Unilever, IKEA, Maersk, and others) together to ramp up corporate support for more stringent climate protection policies by governments. So-called 'Responsible Policy Engagement' is one of the major climate handprint levers for companies, their CEOs and owners.[263] This can either be done directly via public climate action appeals and petitions or through the back door via business associations and direct contact with MPs, mayors and other political players.

Climate action 'bottom up' in companies
What I have described so far applies both on bigger scales

– for example, for medium-sized companies as well as large corporations, but also for the territorial police forces or universities – and on a small scale – for example, for bakeries, hairdressing salons, start-ups or kindergartens; from the executive floor to the warehouse. Of course, the radius of influence differs depending on the position in the organisation: managers and company owners have more power and influence than employees. Nevertheless, everyone can have a much greater impact at work than at home or at the supermarket.

Now, you can of course adopt the handprint perspective individually and apply it in your daily work. This means asking yourself the following question every day and for every process that crosses your desk: How can I contribute to the current decision, process or project so that, as a result, as many people as possible live in a more climate-friendly way than before? Incorporating this question into your daily professional activities is a great starting point.

However, climate action from the middle of a company's workforce is even more effective, if you join forces with colleagues and work together to bring about lasting change. By this I mean, for example, that climate protection is structurally anchored in an industrial company through strict sustainability guidelines for materials purchasing. Then individual climate-conscious employees in the purchasing department (like Julia from chapter 4) don't have to think about climate friendliness every time they place an order. Instead, the sustainable option becomes the standard – for everyone, including the Unwilling and ultimately also for the company's customers.

Get together with colleagues and set up a climate group

You can only join forces if you know who thinks along similar lines. It is therefore extremely important to identify yourself as an environmentalist within the company. In concrete terms, this means not only talking about climate and sustainability in your free time, but also raising the topic with colleagues and superiors during coffee breaks, lunches or team events at work (keeping the 10 tips for fruitful climate conversations in mind). Over time, you will get a good sense of who is also particularly concerned about the escalating climate crisis – and who is motivated to do something about it. You can get together with these like-minded people and set up a committed climate group within the company. Then you have a platform to exchange ideas, demand the points described in the previous section from the company management, keep a close eye on them with regard to their actual priorities, greenwashing and lobbying, support them with implementation and, if necessary, exert internal or public pressure. Incidentally, this is nothing radical or new: more and more people around the world are engaging in outright employee activism to get their employers to take faster and more decisive action against the climate crisis.[264]

One prominent example is the Amazon Employees for Climate Justice group, which has been pushing the tech giant for more sustainability measures since 2018.[265] Initially, a small group led by (now former) Amazon employees Emily Cunningham and Maren Costa held climate talks with managers, but the results were disappointing. The group then published a letter signed by over 8,700 employees to Amazon CEO Jeff Bezos, calling for a more ambitious climate strategy and the transparent publication of company-wide greenhouse gas emissions.[266] The activists also submitted these demands as a draft resolution at the 2019 Annual General Meeting. Although they did not receive enough votes, they gained a lot of attention and support. Ultimately, the group ensured that thousands of Amazon colleagues walked out and took to the streets during the global climate strike in September 2019 – together with hundreds of employees from Google, Microsoft, Facebook, and Twitter.[267] Amid increasing pressure, Jeff Bezos promised net-zero emissions by 2040 and 100 percent renewable energy by 2030.

This great handprint story shows: It is crucial for the success of a climate group to gain influential supporters for its cause. Analyzing exactly where potential and powerful allies exist within or outside the organization makes the group's commitment more effective and success more likely. These can be other colleagues or superiors, suppliers, customers or shareholders, for example, but also politicians, media people or celebrities. It is particularly effective to mobilise trade unions or works councils because they are already professionally structured and experienced at demanding change.

Implementing handprint projects together

If company leadership recognizes the urgency of climate action and has already begun implementing a credible, detailed plan to achieve ambitious sustainability goals – while also advocating for governmental climate policies – employee activism can naturally adopt a softer approach. In this case, the climate group can actively support the company's climate transformation by contributing ideas, handprint projects and specific proposals.

A first important suggestion would be a regular educational program on topics surrounding the broad term of sustainability, i.e. climate and biodiversity crisis, social inequality, just transition, etc. Because only when all interested employees are informed about necessary, possible, just and, above all, effective steps towards true sustainability will the suggestions from the circle of employees extend beyond waste separation and double-sided paper printing to the major and structural causes and levers.[268]

It is best for employees to focus on the biggest causes of greenhouse gas emissions – the three big-point areas that we already know from chapter 6 – that *directly* affect them. This is because in these areas they can formulate change requests in a particularly authentic and comprehensible way. However, climate-friendly structural changes in the work context do not just affect one person alone, but hundreds or thousands of people, depending on the size of the company, because most colleagues behave in accordance with the company's existing rules, social norms, framework conditions and offers.

Similar to private life, *mobility* is also a key issue at work. You can ask yourself: How do I get to work every day? And what can the company do for me and my colleagues to make commuting more climate-friendly? Some ideas would be free annual public transport passes, more carpooling offers, e-charging stations, company e-bikes and a colourful bicycle parking garage with lots of trees instead of company cars with an asphalt car park in front of the company entrance. And for business trips that are unavoidable despite the possibility of video calls or online meetings, employees could receive a company rail card instead of automatically being reimbursed for flight costs by the company.

The second big point relates to *buildings*. The question here is: where and how do I work – and where does the electricity and heating or cooling come from? We already know that every new conventional building generates huge amounts of greenhouse gases because large quantities of steel and cement are used. In addition, every workplace has to be lit, heated in winter and increasingly cooled in summer (thanks to more frequent heat waves). The energy required for this should, of course, come from renewable sources and be produced on site wherever possible, especially through solar panels on all suitable company roofs. Generously greening all other areas and insulating façades provides cooling during hot summers and saves heating energy in the winter.

The third area is *diet*. Here, the question is: What do I eat at work? We know the answer: We should consume plant-based alternatives to animal products and greatly reduce meat consumption. Nowadays, every company canteen should offer tasty vegetarian and vegan dishes alongside a few organic and fair meat or fish alternatives. In addition, measures to reduce food waste have a major impact on the environmental and greenhouse gas balance. For example, I can think of reusable company tupperware boxes, half-price meals the day after or a cooperation with a food bank.

In addition, the *pension fund* investments already mentioned in Chapter 7 are a major influencing factor. This point is rarely considered, although it directly affects each and every employee. After all, their company pensions often flow into more or less climate-damaging capital investments. As with banks and insurers, employees can also demand strict sustainable investment guidelines here – again, ideally with the support of many colleagues, a trade union or the works council.

These are all initial starting points. Of course, there are many other areas that cause considerable amounts of greenhouse gases and opportunities to promote climate action in the company 'bottom up'. They can be developed and tackled step by step together with the climate group and other colleagues as part of regular meetings or climate education formats. The international climate protection organization Project Drawdown has published guidelines for this in 2021, on which many aspects of this chapter are based. I would therefore like to recommend studying their employee guide to Drawdown-aligned business titled *Climate Solutions at Work*.[269]

Chapter 9: Enlarging the Social Handprint

The social handprint is about volunteering in your free time and working with other people to make structural changes that promote environmental behaviour. This can be done at parents' evenings in the nursery, in the football club, in the parish or as a member of a professional climate or environmental organization. Even if there can be overlaps with the political handprint, the social climate handprint is not aimed *directly* at policy changes, but rather via detours – like media, education and community work.[270]

Media and the climate crisis

A lot has happened in the media landscape in terms of reporting on the climate crisis since I started my first climate podcast *Climaware*. At the beginning of 2020, a scientifically sound, easy-to-understand and entertaining climate podcast was still a pioneering achievement in climate communication. Today, there are numerous podcasts, newsletters and sections on the topic.

Nevertheless, the extent of media coverage still does not reflect the relevance and urgency of the issue, nor does it correspond to the population's awareness and need for solid information. In my home country Germany, for example, the climate initiative KLIMA° Vor Acht and the University of Hamburg conducted an analysis including news broadcasts.[271] Their report shows that between 2021 and 2022, only 1 to 2.4 percent of television broadcasting time in the overall programmes of the major channels *Das Erste*, *ZDF* and *WDR* dealt with the climate crisis. For the UK, an analysis by Albert, a BAFTA-owned organisation for environmental sustainability in screen production, showed that as recently as 2020 'cake' was mentioned 10 times more often on British television shows than 'climate change'.[272] Its report was based on analysis of subtitles from almost 400,000 programmes from major TV channels, but excluded news broadcasts. Consequently, only half of the people in countries like France, Germany, the UK, and the United States see, read, or hear news or information about climate change on a weekly basis. [273]

And *if* the climate crisis is discussed on television or in newspapers, there is often misinformation or disinformation. In 2024, 17 percent of Britons stated they are regularly exposed to false or misleading climate information.[274] Moreover, climate coverage is often too passive, too populist, too incomprehensible or too abstract. This is the conclusion reached by 28 authors from the fields of journalism and communication science, among others, in the book *Media in the Climate Crisis* in 2022.[275]

Fortunately, Psychologists For Future's climate media guide shows how things can be done better and makes three key recommendations. First, the climate crisis must be considered highly relevant and consistently covered by all major TV channels, newspapers and online media, ideally in prime time and on front pages. Secondly, emotions should be taken up as an appropriate reaction to the reporting. And thirdly, the media should highlight constructive solutions, coping mechanisms and possible courses of action, ideally with positive examples from near and far.[276]

It is legitimate to put pressure on the leading media to report and educate people about the climate crisis in line with its relevance and society's need for information. On the one hand, this can be done by supporting initiatives such as the Oxford Climate Journalism Network[277]. On the other hand, you can also contact journalism associations, publishing groups or TV stations directly and write constructive and critical open letters to them, for example. Most large media companies take these letters seriously and many newspapers even print them occasionally.

Beyond that, of course, you can simply take matters into your own hands. Even if you probably won't immediately achieve the public reach of Leonardo DiCaprio, you can still try to gain a certain reach beyond your friends and family – and use it for climate education, your own climate action experiences and messages that motivate others to take action. Who knows: perhaps there will be one or two new climate communication talents among the readers of this book.

Demand and promote climate education at all educational institutions

Of course, climate education should not only take place more on television, radio, in newspapers or on the media platforms of celebrities and influencers, but should also be given more space in all educational institutions. By this I mean above all schools and colleges and so on. Addressing the climate crisis must be an integral and mandatory part of every official curriculum and training program. However, without presenting our all-encompassing ecological crisis as a problem of individual consumer decisions, as has usually been the case to date.[278]

Of course, the ecological reality can and must also be addressed more in municipalities, museums, cultural centers, theaters and the arts.[279] Because the climate crisis touches, influences or revolutionizes pretty much every discipline, every education, every job, *every part of life*. And large-scale, politicizing climate education programmes set the course for managing the turmoil of the decades ahead.[280]

If local and regional government and education ministries are reluctant to incorporate climate education programs into teacher training, examination regulations and curricula, a little pressure and involvement on the part of students and parents can certainly do no harm. Students can approach their teachers and class representatives directly or join forces in the student council. Parents can actively campaign for climate education in schools for the benefit of their children's future.

At least those responsible at the University of Barcelona seem to have realized that climate education is not simply 'nice to have' and should only be offered in voluntary lecture series, but that the climate crisis is an overarching issue that affects everyone. After students occupied parts of the university campus for a week in protest in November 2022 and around 200 Spanish professors supported their demands, the university agreed: all students will have to complete a compulsory course on the climate crisis. In addition, all of the university's 6,000 academic staff are to undergo a training program on climate issues.[281]

However, the University of Barcelona did not implement one of the students' central demands: to prevent financial contributions from oil companies such as Repsol and cooperation with the fossil fuel industry. Research and donations from companies as well as the university's own capital investments represent an influential lever that students and university staff should not ignore.[282] Indeed, campaigns have successfully ratcheted up the pressure on British institutions to withdraw their investments in oil companies. In 2020, Oxford University relented, announcing it was divesting its endowments from the fossil fuel industry and asking fund managers working on its behalf to show their plans for reaching net zero carbon.[283] And in 2024, Cambridge University adopted a moratorium on grants and philanthropic bequests from fossil fuel companies after a campaign from students and academics.[284] We see that a mandatory climate course like at the University of Barcelona is not enough. Instead, all educational institutions need to be holistically adapted to the challenges of the future, including operations, management, finances and so on.[285]

The key question is: What skills are essential in a world warming by probably another half a degree – one that must function without coal, oil, and fossil gas? Should children in nurseries and primary schools still be taught that a healthy diet requires meat on the plate and cow's milk in the cup? Should secondary school pupils still learn that Gross Domestic Product (GDP) is the ultimate measure of progress, success, and well-being? And should education ever truly be 'finished' – or is lifelong learning not far more important?

Start a voluntary collective climate commitment

The German TV moderator, entertainer and climate philanthropist Eckart von Hirschhausen said something to me during our podcast recording that stuck with me: 'The most important thing an individual can do is not to remain an individual.'[286] So, similar to the climate group at work, you should also get together with like-minded people for collective climate action in your free time.

One option is to look for an overarching climate initiative that interests you, is transformative and whose values and concerns you share, and then to found a local branch of this initiative. Of course, you can simply find inspiration from the many For Future groups. Or, during a personal climate lesson in a climate podcast or newsletter, you may come across a national or regional climate protection campaign that you would like to establish or support locally.[287]

Another option, which requires a little more creativity, is to launch a completely new climate initiative together with a few fellow campaigners. Of course, it makes sense to focus on topics that are particularly close to your heart and where you ideally already have some expertise and experience. But ultimately, creativity can be limitless. It doesn't matter whether it's solidarity agriculture or urban gardening, an app for cycling miles against friends, a neighbourhood guide for the most climate-friendly street parties, a cinema initiative that organizes film evenings on climate issues, an informal climate after-work meeting or a climate hike across Europe.

If you can't come up with a good idea for a new climate initiative or your own project, you can simply register for the next 'hackathon' for the climate. In these digital ideas and implementation crash courses, you and other participants have 24 or 48 hours to come up with an idea that is as effective, inspiring and feasible as possible to advance climate action in society. If you convince the jury, you can implement it immediately with the prize money you win.

Actively support a professional environmental organisation

However, it is easier, quicker and often more effective not to set up a new climate initiative or organisation, but to join an existing one. Of course, this option also depends on what you want to do. On the other hand, the initiative or organisation should also meet certain criteria in order to have the greatest possible impact.[288] Among other things, it should have shared values, a positive vision, clear goals, professional management structures including clear decision-making processes and responsibilities and, ideally, sufficient financial and human resources. Ideally, it should also have a transformative, far-reaching solution strategy ('theory of change') that can be scaled up and thus contribute to the global climate transition. If the organisation works in this way to reduce greenhouse gases on a large scale – or helps as many people as possible to live in a more climate-friendly way, preferably through structural changes – then it is on the right track.

One thing is certain: without the involvement of hundreds of volunteers, there would not be all the for-future groups and their climate action successes of recent years. And professional climate associations and environmental organisations – such as Greenpeace, WWF, Friends of the Earth and many more – are also dependent on volunteers. The right place to significantly increase your own social climate handprint is often just a few emails, phone calls or participation in an onboarding evening away.

Climate commitment in voluntary work, leisure associations, clubs and parishes

If you don't want to or can't take on a new voluntary position for environmental protection, you still have the opportunity to increase your social climate handprint. This alternative route is via the social networks and communities that you are already part of in your free time.

By this I mean comparatively loose and informal groups – such as the aforementioned parent-teacher evenings in schools, the neighbourhood or a book club. Here, the ideas for joint climate action campaigns can range from climate fundraising runs and the demand for a PV system on the school building to energy-saving and cycling competitions among neighbours to a climate nonfiction reading month.

On the other hand, I am also referring in particular to formal, more established community structures such as associations or church congregations, whose task is not primarily climate or environmental protection, but in which you can still work effectively for these causes. I would like to take a closer look at these more established community structures, as these pillars of our society have so far been used far too little for collective climate action.

Clubs

Just how great the potential influence of the world of associations is becomes clear when you consider the following figures: approximately 1 million British people volunteer to help out organisations such as sports and hobby clubs on a regular basis, almost one quarter of the population.[289] If even just some of these activities were geared more towards sustainability, this would have a considerable effect on accelerating the climate transition.

And individuals can also increase their own handprint by helping to reduce the footprint of the club. Measures such as thermally insulated changing rooms, PV systems on club buildings, heat pumps for heating sports halls, more efficient stadium lighting or e-charging and bicycle parking facilities not only create climate-friendly club structures, but are usually also generously subsidized.[290]

However, the (sports) club as a whole can also increase its climate handprint, similar to companies (see chapter 8). With a little creativity and motivation, all these clubs can organize the first steps on the way to changing structures, co-ordinating second-hand clothing exchanges, food festivals with edible food from food waste, fundraising concerts for climate protection projects, dance festivals between wind turbines, bike tours along main roads, declaring climate action match days and use half-time or breaks for climate education, and many other ideas. As mentioned before, it is also important to think about the association's account: Is it possible to switch to a more climate-friendly bank?

I would particularly like to emphasize charity clubs such as Lions, Rotary or Round Table at this point, because these clubs usually bring together people of above-average wealth to cultivate friendships and do good. Awakening or strengthening their commitment to climate and environmental protection and aligning the service and fundraising activities of these clubs more closely with the fight for a habitable planet will therefore have a major impact. In addition to supporting social projects that help people directly, it is now more important than ever to support transformative climate solutions so that many more people do not have to be helped in the future. As a Rotary club, for example, this means not only carrying out a fundraising campaign for the food bank, but also one for a climate organisation so that the grains for the food banks continue to grow in the future.

Parish

Churches, mosques, synagogues and other religious organisations can also work fervently for the preservation of creation, peace and the fight against hunger and poverty in the world by using all their resources to rebel against the environmental and climate emergency.

Since Pope Francis published his climate and environmental protection encyclical *Laudato Si'* in 2015, the climate issue has fortunately gained momentum in the Catholic Church.[291] The Church of England has set out a roadmap for becoming zero carbon by 2030.[292] Despite such encouraging steps, there is still a lot of untapped climate action potential in religious organisations.[293]

As in companies, schools and universities, all members of a congregation can demand and implement this 'from below', so to speak.[294] If you are a worshipper, you can support the following handprint ideas and actions in the church, among others: a part of the collection goes to climate action projects and organisations, sermons call more often for active climate commitment, the church, in keeping with its Christian tradition, once again makes a stronger commitment to veggie Fridays,[295] the next choir trips or church camps take place under the motto 'climate education and action', the churches publicly show solidarity with climate activists and demand stricter climate policies from the local or national government, the church assets are invested in a climate-friendly way.[296]

No time for voluntary work? Handprint on the side and in between

Yes, you need time for all the possibilities and ways I have described so far to accelerate and actively shape the urgently needed climate transition in initiatives, organisations and community structures. In any case, time is one of our most important personal resources in the fight against the climate crisis. But what if you don't have time? Can even a permanently busy person increase their social climate handprint?

First of all, I would like to invite all those who think they don't have time for climate action to do some honest self-reflection, especially the more privileged among us: Is this really true? Do we really have no time – or is it not rather the case that we are simply prioritizing other things and activities? If you think about your most important values in life, then consider how these are threatened by climate chaos and a potential collapse of our ecological, social, political and economic systems, and then list the activities which you spend roughly one hundred waking hours a week doing, you may recognize a discrepancy. That would be a sign to make contributing to saving the planet a new priority in your life. And it would be an opportunity to ultimately live a more value-based, less self-alienated and more content life.

For those who, even after such honest self-questioning, still come to the conclusion that they have no time for active climate or environment protection commitment, I would nevertheless like to present two ways of increasing the social climate handprint: firstly through donations and secondly through 'customer activism'.

Donations

The first option in particular may be a good option for busy people. After all, once a monthly donation membership has been set up within a few minutes, it repeats itself and continues to increase our climate handprint month after month without us having to think about it again and again. (This is particularly aimed at those of us who tend to be wealthy – especially if that wealth was inherited or acquired with less work compared to other people).[297]

Felicitas von Peters, founder and managing director of Active Philanthropy, supports foundations and donors who want to get involved in climate and environmental protection with learning trips, online tools, starter kits and advice on funding strategies and management, among other things.[298]

Of course, donations alone cannot cover the approximately *8 trillion US dollars* that the United Nations Environment Programme (UNEP) estimates will be needed by 2050 to tackle the intertwined climate and biodiversity crisis.[299] But according to Felicitas von Peters, climate philanthropy can be a powerful catalyst for this. Especially with projects that aim to achieve as much impact as possible per pound donated.[300]

This is precisely the strategy of 'effective altruism' and could also be a very helpful approach if we want to donate to climate and environment protection[301]: the greatest possible impact per pound donated. In addition to the well-known groups I described earlier, such as WWF, Greenpeace, Friends of the Earth and others, we should also focus on protest movements. According to analyses by the Social Change Lab, these can be far more effective than traditional environmental and climate protection organisations.[302] (Incidentally, this is where the social climate handprint overlaps the *political* climate handprint). In 2019, for every British pound donated to Extinction Rebellion (XR) in the UK, around 12.5 tons of greenhouse gases are said to have been reduced (due to the climate policy adjustments ultimately brought about by the protests).[303] The Clean Air Task Force organisation, which campaigns for political and technological change in the USA, has also been shown to reduce greenhouse gas emissions quite effectively: for every US dollar donated, this organization reduces greenhouse gas emissions by at least 1 tonne.[304]

If you want to increase your climate handprint with monetary donations, you should therefore not just listen to your gut feeling but rather research effective donation recipients yourself. For example, you can find out more from the Social Change Lab, Founders Pledge or Giving Green.[305] For my part, I regularly donate to the Climate Emergency Fund, which supports climate activism groups in various countries, as well as to traditional environmental NGOs.[306]

The customer is king

The second way to advocate for climate-friendly structures and offers in everyday life without having to make major time commitments is to repeatedly advocate for climate-friendly products and services as a customer with verbal and written feedback. Such 'customer activism' therefore goes beyond simply buying climate-friendly products. Rather, it involves direct feedback – alone or, even better, together with others – to ensure that even more people can purchase or use climate-friendly products and services as easily and cheaply as possible in the future. This can be done, for example, by pointing out existing obstacles and the lack of climate-friendly options to suppliers and companies or by demanding or proposing climate-friendly alternatives.

In my experience, the best strategy is to fill out a customer feedback form or send an email to the service team and then follow up a few days later in person or by calling the staff. For example, I repeatedly pointed out to my gym provider that there was no bicycle parking near the entrance and that the TV screens in the gym were advertising private jet travel. A few days later, I actually received a call back from the manager: they were already lobbying the city to build bicycle parking spaces in the entrance area and would discuss a ban on advertising for particularly climate-damaging providers such as private jet rental at the next meeting. (At the time of writing, there are still no bicycle parking spaces, but at least I no longer have to look at private jet advertisements when I'm doing my workouts).

Anyone who goes through the world with climate awareness and open eyes will always and everywhere come across opportunities to use their customers' voices to initiate change. Whether it's asking at hotel receptions why there is no PV system on the roof or how the heating is done, friendly complaints in restaurants where the only vegan options are unimaginative salads, or tips in the supermarket that you would like to buy vegan minced meat. After all, economics is not about 'supply and purchase', but 'supply and demand'. This is because demanding can change supply. This is especially true if you demand repeatedly and collectively.

Chapter 10: Enlarging the Political Handprint

The political arena is where it becomes clear whether we as a society are prepared to demand, adopt and accept the policies and measures that will preserve our precious planet for us and our children – and improve our health and energy. Because in politics, our elected representatives set the rules of the game for our society. They decide on our behalf what we cannot decide on our own: For example, whether short-haul flights and beef will become more expensive than train travel and organic vegan food due to the carbon pricing, whether low-income families with their low carbon emissions will receive an annual climate bonus from carbon revenues, whether people will still be allowed to drive at high speeds on motorways or install new natural gas heating systems, where cycle paths, railroad lines and wind turbines will be built, how much money will be invested in the electricity grid, how many wetlands will be restored, how many forest will be reforested and how much land will be renatured. In short, politicians can resolve the stubborn structural obstacles to truly climate-friendly living and initiate the necessary measures to adapt to the unavoidable consequences of climate change.

However, politicians are elected to represent us. So politics is us, each and every one of us. This means that the political sphere offers us particularly effective levers of influence. These are primarily our votes and democratic rights, such as legal action, political co-determination and influence as well as the collective power of activism and protest movements.

We have the choice, and the rights!
'What is the first civic duty, Gabriel?', my father often asked

me when I was little. 'Keep calm,' I always replied – without really knowing what that meant. 'And the second civic duty?' my father followed up. And I answered again automatically: 'Vote.'

Today – in this planetary climate emergency – I would give my father a slightly different answer. When asked about our first civic duty, I would answer: 'Keep calm – but get involved, without panic, but with priority.' And my second answer would be: 'Vote – but not just for myself, but also for my ancestors who fought for this right and for my descendants who will have to live with today's political decisions.'

Because the fact is: yes, our governments have a special responsibility to set the framework conditions for the economy and our coexistence in such a way that we achieve our legally agreed and existentially important environmental and climate targets. However, government politicians are always looking at opinion polls and the next election because their political careers depend on our favour – their voters. So ultimately, the special responsibility for curbing the heating and destruction of our environment falls to us again. And that means voting for and supporting parties and politicians who really take the climate crisis seriously – and don't just pretend to.

Making every election a climate election

The carbon clock for limiting global warming to 1.5°C is ticking relentlessly. At current greenhouse gas emission rates, our global carbon budget for staying within the 1.5°C threshold is expected to be exhausted by the end of this decade.[307] A legislative period usually lasts four to five years. Planning, approvals and the construction of new infrastructure, such as wind turbines, hydrogen pipelines, e-charging stations and the like, can easily take just as long. This means that the next political elections will decide whether we can still achieve our optimistic climate target. So from now on, every election is a climate election – regardless of whether it is a mayoral, local, state or federal election.

However, when it comes to voting for effective climate action, the devil is in the detail. The questions that voters should definitely keep in mind when reading election manifestos or following the election campaign are therefore: How much detail do the parties and politicians go into? Do their proposals and ideas go beyond clichés and buzzwords such as 'technological solutions' and 'carbon capture and storage' (CCS) and hype topics such as 'e-fuels' or 'green hydrogen'? How do climate scientists and experts assess the parties' election manifestos? Do they really contain systemic and transformative concepts? Or are they just smoke and mirrors to promise a certain clientele that things will continue as they are without any far-reaching changes?

Claiming the right to a sustainable environment and climate

However, we all not only have the choice. We also have the right. In 2022, the United Nations General Assembly recognized the right to a clean, healthy and sustainable environment as a universal human right.[308] And in my home country, for example, the constitution of the Federal Republic of Germany states: 'The state shall protect the natural foundations of life and animals within the framework of the constitutional order, also in responsibility for future generations [...].'[309] Effective climate protection has constitutional status in many countries and can no longer be postponed. And if governments postpone climate measures and policies, courts will be breathing down their necks.

Since the Paris Climate Agreement was adopted in 2015, the number of climate court cases worldwide has more than doubled, with a quarter of all climate lawsuits ever filed in recent years.[310] Court proceedings on the climate crisis are now taking place at all levels. For example at the International Criminal Court, which the island state of Vanuatu activated by means of a resolution on climate issues,[311] and at the European Court of Human Rights, before which groups of young people have been taking legal action against the inadequate climate policy of some European states since 2022.[312]

Companies in the fossil fuel industry are also increasingly confronted with climate court cases. These involve lawsuits about emission reductions, compensation for damages and fraud, as in many lawsuits against 'Big Oil' by US states, cities and individuals.[313] For instance, ExxonMobil's in-house scientists predicted current global warming fairly accurately back in the late 1970s, but the company's management chose to engage in public climate denial and targeted disinformation campaigns instead of transforming the business model.[314] Furthermore, these climate lawsuits are about protection from climate damage, as in the case of Peruvian smallholder Saúl Luciano Lliuya. Since 2015, he has been demanding proportional protection payments from the German energy company RWE against the threat of flooding caused by the melting of glaciers in the Andes, as RWE contributed to this through its greenhouse gas emissions.[315]

The many successful lawsuits in recent years show this: Legal action is a powerful lever that we can use to achieve a large climate handprint. Although this path is often time-consuming and sometimes nerve-wracking, you don't have to undertake it alone. On the one hand, we can count on the support of experienced organisations such as ClientEarth, Greenpeace, and Friends of the Earth. On the other hand, there are now many experienced lawyers who support climate lawsuits professionally and on a voluntary basis.

Quiet, loud or uncomfortable: helping politics on its feet
Words, promises, climate targets and laws alone are no guarantee for actual climate policy measures and meaningful environmental protection. Otherwise there would be no need for all the climate lawsuits. So we all need to be vigilant and

keep a close eye on our politicians as they work to preserve our livelihoods – and help them with legal remedies if necessary. However, it is not only in individual moments such as elections or legal climate rulings that we citizens can influence politics. There are also many ways to get involved in the ongoing political process, to help political decision-makers with a little pressure and to push through environmental concerns.

In general, there are three peaceful ways to do this: The first way is the usual democratic procedure of bringing about change through the political institutions – for example, by joining a political party or petitioning parliament. The second type is the use of legal means of protest, such as taking part in a demonstration. And the third type is forms of protest from the spectrum of non-violent civil disobedience. Some people refer to this as 'radical' protest because the actions are deliberately disruptive of everyday life and against the rules, such as protests by Just Stop Oil in the UK.

Political commitment

There are more than 1,700 political representatives in the UK Westminster parliament and national assemblies; in the US 530 members sit in the two houses of Congress. Then there are all the local and honorary politicians. Together, they form the heart of our democracies. Joining this core as a politician naturally gives you an extremely large handprint potential.

Of course, only few people have the will and capacity to become full-time politicians. But you can also significantly increase your climate handprint by getting involved in politics alongside your job – for example as an honorary member of a local or city council, where you can have a say in many climate-related decisions. If you are not able to devote the necessary time to this, you can at least join a democratic party. Here, every member is free to decide how much time they want to spend on (climate) discussions, meetings, election campaigns and other party work. One factor that should not be underestimated is the right to take part in party conferences, member surveys and votes – especially when it comes to the next election program or a government coalition agreement.

Petitions, campaigns and municipal climate action
However, democracy is not just about voters and active politicians. No, there are many other ways to increase your political handprint: Instead of exerting direct influence on political processes and decisions within party headquarters, town halls and parliaments, it is also possible to do so from the outside – with online petitions, applications to the city administration, emails to mayors or MPs and much more.

As a result of pressure from voters, more than 600 councils in the UK, including London, have declared a climate emergency, meaning that they have committed themselves to acting urgently to safeguard the planet. It is estimated that these actions will save 2.5 billion tonnes of greenhouse gases.[316] Similarly, the Transition Towns movement, which started in the UK with the aim of increasing self-sufficiency without fossil fuels, has now spread to almost 1,000 communities worldwide.[317]

Citizen lobbying for climate and environment protection
If you have concrete ideas for climate or environmental protection in your area, you should definitely approach your local politicians (repeatedly) and persistently push for their implementation. But even without specific proposals or project ideas, it helps to approach politicians about the climate issue in general. The email addresses and telephone numbers of local councillors and the member of parliament for your constituency are publicly available on the internet. The more we let them all know that we care about a rapid and

just climate transition and that they should trust us to take the necessary measures, the more likely they are to support it. '[Because] nothing scares a ruling party as much as a clearly expressed displeasure from the population or a clearly expressed wish,' said Sylvia Kotting-Uhl, former chairwoman of the German Parliament's Environment Committee, on the Climaware podcast at the end of her more than 30-year political career.[318]

Making your voice heard, saying what you want – in a way, that's lobbying. And that's exactly what it's all about: more *citizens lobbying* for fast and effective climate action, as a kind of counterweight to the fossil fuel and status quo lobbying of some corporations and their influential networks. For example, over 1,700 lobbyists from coal, oil and gas companies were registered for the 2024 UN Climate Change Conference 'COP29' in Azerbaijan.[319]

On the one hand, citizen lobbying for climate and environment protection can be done by letter, email or phone call. However, face-to-face meetings are more effective. This is indicated by the results of a Canadian study from 2021 as well as testimonials from climate activists.[320] At this point, I would like to emphasize once again the climate protection influence potential of established (business) associations and trade unions, to which politicians usually devote even more attention and time than to individuals and loose initiatives.

Another relatively new way for citizens to make their voices heard by politicians is through citizens' councils. For such a council, citizens are drawn by lot using a special method so that the council members reflect society in terms of characteristics such as age, gender, school-leaving qualifications and migration history. They meet over several dates and receive information and support from various experts on a clear question. In moderated small groups, in which everyone has an equal say, the members then develop recommendations for policymakers. Remarkably, the results from two experimental citizens' climate councils on a national level in Germany and Austria were significantly more ambitious in terms of climate protection than the actions of the national governments.[321]

One last item should not be missing from the lobbying list: *money*. Everyone can donate to parties that take the climate emergency seriously – especially during election campaigns, when every vote counts.

Demonstrations and protest

In democracies, majorities are known to determine change processes. However, transformation research shows that the prevailing majority opinion can usually be changed through the commitment of an engaged minority. The latest IPCC report, for example, states that between 10 and 30 percent of committed individuals are needed to set new social norms.[322] Public and peaceful protests are particularly effective when it comes to influencing the majority opinion of the public and enforcing political demands. A study of protests from 1900 to 2006 shows that almost every government – even in dictatorships – has had to give in at least partially to people's demands when a small but committed minority of the population repeatedly and peacefully took to the streets.[323]

This is also demonstrated by the mass climate protests triggered by Greta Thunberg and Fridays For Future after 2018, which applied the pressure for the EU Green Deal, the coal phase-out and the 'Climate Protection Act' in Germany and putting the climate crisis high on the agenda in following elections.

'Politics is what is possible,' said German Chancellor Angela Merkel at the time.[324] Yes, but demonstrations and pressure from the population expand the scope of what is politically possible. That's probably what EU Commission President Ursula von der Leyen meant when she replied to my question about her advice to the younger generation: 'Keep at it, don't be put off by the fact that some processes seem tough, we can do more, there's still a lot in it, and we need you and the pressure you exert.'[325] Because then politicians will have a clear democratic mandate for what is needed: to introduce truly far-reaching, transformative and socially just climate protection measures now.[326] Politics is therefore not only what is possible, but also what becomes possible because people make their voices heard.

I myself have taken part in all global and many local climate strikes organised by Fridays For Future. Not because I agree one hundred percent with every slogan or would make friends with every person at the demonstrations. But because the principles of the movement are right. And because I can always feel the powerful sense of togetherness that arises when you stand up loudly and publicly for your values as part of a crowd, as an antidote to individual powerlessness and climate anxiety. However, my most formative experience of the power and influence of Greta Thunberg and Fridays For Future came quite unexpectedly on December 6, 2019, when I was at the huge IFEMA exhibition center in Madrid during the UN Climate Change Conference 'COP25'.

More than just a spark of hope

The staff at the UN Climate Change Secretariat were working hard, even though the interim results of the first week of negotiations were sobering. I was working on invitation lists for an event when a colleague to my right suddenly jumped up after glancing at her smartphone and shouted: 'Greta is here!' Without hesitation, everyone in the room literally dropped their pens and ran after the running colleague.

When I arrived in the airport-sized reception hall a few minutes later, I saw that we were not the only ones waiting for Greta Thunberg's arrival. A larger crowd had formed around the teenager than for any previous head of state, minister or celebrity. Some of the otherwise quite discreet and reserved negotiating delegates tried to get a photo with her like groupies in a boy band. And within a few minutes, a hundred-metre-long queue of journalists and onlookers had formed in front of the entrance to the press conference room. My colleague cursed under her breath, but returned to her desk with a hopeful smile.

The mood at the conference changed abruptly. The fact that the climate figurehead of the younger generation had made the arduous journey across the Atlantic in time gave everyone I spoke to new courage and fresh impetus. And just a few hours later, the spark of hope had ignited the whole city: That very evening, more than 100,000 people gathered in the city centre for Madrid's biggest climate demonstration ever. Looking at the screens of their smartphones, all politicians [327]suddenly realized: 'They are watching us.'

The air in the conference halls was literally vibrating, I could feel the pressure. The fact that the sluggish negotiations on a global carbon market were ultimately postponed to the next 'COP26' in Glasgow instead of opening the door to global greenwashing with a lousy set of rules is, in my opinion, largely thanks to Greta Thunberg, Fridays For Future and all the people who took to the streets of Madrid on December 6, 2019.

Even though the coronavirus pandemic and its lockdowns took the wind out of Fridays For Future's sails, there is still a lot of energy and potential in the climate movement. We should not underestimate the influence of public climate protests. The former head of the UN Climate Change Secretariat, Patricia Espinosa, told me that climate demonstrations by young people were her greatest source of hope.

For many people, demonstrating is something unfamiliar. They feel uncomfortable at the thought of marching through the streets with strangers and their banners, flags and cardboard signs, singing loudly, drumming and chanting. I felt the same way before I took part in my first demonstration.

But the inspiring feeling of unity can of course only come to those who try it out for themselves. Maybe the first time without a sign and whistle and with a friend by your side, just as a walk. My partner and I have already taken a few demo newcomers to global climate strikes and usually got the following feedback afterwards: 'It wasn't bad at all. On the contrary – I actually enjoyed it. More often from now on!'

Nevertheless, not everyone wants to publicly stand up for their values in a loud way. And that's perfectly okay. But then you can ask yourself in what other places and at what other times you can demonstrate or protest. Perhaps you've already had the feeling in a certain situation that you need to say something because things can't go on as they are: at a meeting with the company board, at a general meeting, with friends or at other events. Having the courage to stand up for your values and convictions more often, to sometimes do something uncomfortable, can also increase your handprint.

Civil disobedience and other non-violent ways to protest

Someone who stood up again and again throughout his life and always said what he didn't like was the US-American Henry David Thoreau. For his contemporaries, he was an uncomfortably stubborn man, but one with high moral principles. In 1846, he spent a night in prison for refusing to pay his taxes to the US government, thereby supporting slavery and the US war against Mexico. 'If you are to be forced to participate in injustice against another, then break the law!' he later wrote in an essay whose title would forever coin the term for this attitude and the actions that followed: civil disobedience.[328]

This is the third level of protest that can be used to make a major political impact. It is part of a liberal, constitutional democracy, argue well-known philosophers such as John Rawls and Jürgen Habermas.[329] This is because only individual rules and laws are violated – deliberately, publicly and peacefully – in order to draw attention to a serious moral injustice or acute social grievances. Otherwise, civil disobedience activists respect the legal framework or the democratic constitution as a whole. They therefore accept the legal consequences of their often disruptive direct actions with approval and without resistance (at most with passive resistance).

In this way, some major ethical advances have been achieved since Henry David Thoreau's night in prison.[330] The so-called 'suffragettes' fought for women's suffrage in Great Britain by means of civil disobedience. Both Mahatma Gandhi and Martin Luther King Jr. were inspired by Henry David Thoreau to create the successful protest forms of the independence movement in India and the anti-racist civil rights movement in the USA. Nelson Mandela also used such methods to lead South Africa out of apartheid. Used correctly, this form of protest is therefore quite effective in initiating desirable changes in society. What's more, according to German protest researcher Felix Anderl, there is a consensus in his discipline 'that nothing changes at all without crossing the line'.[331]

Of course, even the noblest end does not justify all means. Caution and sensitivity to the *limits* of crossing the line are absolutely essential in civil disobedience. And such a limit is clearly crossed when violence is used against people. (Although it is debatable how broadly the concept of violence should be defined: does this also include damage to property? Is blocking a road a form of violence?) Beyond that, violence is pointless anyway: Peaceful forms of protest are demonstrably about twice as successful as violent ones.[332]

The big fuss about Just Stop Oil
The 'radical' parts of the climate movement also use civil disobedience to shake up society and politicians. For example, in 2019, activists from Extinction Rebellion (XR) climbed onto

London subway trains with protest banners during rush hour and staged sit-in blockades at road junctions in many cities. And in 2022 and 2023, the activism group Letzte Generation ('Last Generation') dominated the German-language headlines. Their name comes from a quote from former US President Barack Obama: 'We are the first generation to feel the impact of climate change and the last generation that can do something about it.'[333]

And in Great Britain, Just Stop Oil made it to the media front pages by blocking motorways and carrying out protest stunts such as throwing soup at the safety glass of a painting by Vincent van Gogh at the National Gallery in London.[334] This brought huge attention and just as much indignation.[335] This is understandable: nobody likes to be stuck in a traffic jam on the way to work because a bunch of young people in high-visibility vests delaying their journey by hours – no matter how noble their motives.

At this point, however, we should be aware of the goal of this kind of protest. This is not a majority of popularity in opinion polls, but the greatest possible media and public attention for the failure to meet our legally binding climate targets in order to ultimately exert maximum pressure on political decision-makers. Pushing the boundaries of what can be thought, said and done about the climate transition and making politically acceptable what was previously unimaginable. According to the two British climate communication experts Adam Corner and Jamie Clarke, successful 'radical' protest opens up new political space into which the mainstream can follow.[336]

A certain amount of initial public outrage is therefore completely normal – indeed, it is even desirable in order to put the moral cause in the media spotlight. However, it is important for the overall positive impact that this initial outrage subsides at some point and turns into broader support over time, at least for the more moderate parts of the movement.[337] The Fridays For Future school strikes, which were initially condemned and defamed, but were later praised by the majority, showed that this can work.[338]

Another, particularly drastic example is an opinion poll of the US population from 1961, which concerned the sit-in blockades of the so-called 'Freedom Riders' and their banned 'mixed race' bus trips to the southern states: At the time, a clear majority in the US rejected the Freedom Riders' protest methods, believing they did more harm than good to the causes of the Black US citizens.[339] Today, of course, it's the other way around: the majority of people view their civil disobedience positively.

Who knows how the British protesters taking dramatic action for the climate will be viewed in the future? Surely, almost certainly better than now. New repressive laws passed in the UK have led to more and more climate protesters being arrested and given lengthy prison sentences. The two activists who threw soup at the security glass in front of one of Van Gogh's paintings in the National Gallery London in 2022, for example, have to spend 20 and 24 months in prison.[340] Research found in 2024 that climate protesters in the UK are arrested at three times the global average rate.[341] At the end of 2024, nineteen people were in prison after taking part in Just Stop Oil's protests.[342] Michel Forst, the UN special rapporteur for environmental defenders, described the situation in the UK as 'terrifying' and warned: 'In many countries, the state response to peaceful environmental protest is increasingly to repress rather than to enable and protect those seeking to speak up for the environment.'[343]

Is that legitimate?

A look at history shows that it is important to distinguish between legality and legitimacy. Not everything that is legal is morally legitimate, and not everything that is legitimate is legal. I understand this sentence against the backdrop of human overexploitation of nature, which is still legal in many ways but illegitimate, the sixth major mass extinction in the history of the earth and catastrophic climate change. With this in mind, the question arises: is civil disobedience a legitimate form of protest against the climate crisis?

The former head of the UN Climate Change Secretariat, Christiana Figueres, says yes: 'It is time to participate in non-violent political movements wherever possible. [...] Civil disobedience is not only a moral choice, it is also the most effective way to influence world politics.'[344] More and more voices from the scientific community are also joining this assessment.[345] In April 2022, for example, over 1,000 scientists worldwide engaged in civil disobedience under the banner of Scientists Rebellion.[346]

In conclusion, I would like to say: Yes, civil disobedience divides opinion.[347] But it belongs on the list of political handprint levers. Because with the right protest strategy and attitude, this form of protest is both a legitimate and effective means of advancing the social climate transition.[348] According to protest research, it is even a necessary one given the scale and speed of the transformation we are facing.

However, the various types of non-violent protest actions are difficult to assess individually. Whether throwing soup or glueing doors together are suitable for this purpose will only be answered in retrospect in a few years' time. The overall effect on society is not (yet) clear.[349] Groups like Extinction Rebellion (XR), 'Last Generation' and Just Stop Oil have experimented with various forms of civil disobedience since 2019, and of course you don't have to approve of every stunt or action. But in my view, it would be negligent *not* to try such experiments during this crucial phase of the planetary climate emergency.

Epilogue: Beyond the Handprint

According to Michael E. Mann, one of the most renowned climate researchers in the US, containing the climate crisis in time would require a level of social mobilisation comparable to that seen in the US after Japan's attack on Pearl Harbor and its subsequent entry into the Second World War.[350] At that time, the government was suddenly able to take rapid and drastic measures, such as restructuring the economy and introducing the broadest and most progressive tax in US history – the 'Victory Tax' – which included a wealth tax of up to 75 percent for multi-millionaires. This illustrates that, in addition to dangerous climate tipping points, there are also social tipping points – moments when societies can shift relatively quickly in a particular direction.[351]

Tipping into sustainability with a social tipping point

Together with participating scientists from the Potsdam Institute for Climate Impact Research, transformation researcher Illona Otto identified a number of social drivers that can be used to trigger positive social tipping points to accelerate the climate transition.[352] According to their 2020 study, the main points are as follows: Spreading climate education; highlighting the moral consequences of burning fossil fuels; measuring and publicly disclosing greenhouse gas emissions everywhere; ending all subsidies of and investments in fossil fuels; scaling up decentralised and renewable energy production; and making cities climate-neutral. Experts from the transformation consultancy Systemiq even describe a possible cascade of positive tipping points if our governments give the right impetus now.[353]

For me personally, the concept of positive social tipping points is a source of hope: it shows that a rapid climate turnaround is not only technically and economically feasible, but also socially possible. It shows that we can wake up from our previous collective handprint oblivion to finally want, do and politically demand what is necessary as a society. And it shows that it depends on each and every one of us who is open-minded, engaged, and willing to activate handprint levers. Because every single handprint increase could be the one that triggers another positive social tipping point for true sustainability.

Please don't misunderstand me: I am not trying to promote excessive or naïve optimism. The transformation ahead will not be easy. We are facing the greatest challenge humanity has ever encountered – especially considering the other crises deeply intertwined with the climate crisis: the threat of biodiversity collapse, severe environmental pollution, extreme inequality, and the growing concentration of wealth, media, and political power in the hands of a few ultra-rich, posing a serious threat to our liberal democracies.

However, if we navigate the fine line of stoic action – avoiding the twin traps of deluded wishful thinking on one side and fatalistic doomism on the other – we can achieve this transformation. And it will be worth it! Because on the path to climate neutrality, we will build a healthier, fairer, and better world.

In order to successfully tread this path – the narrow ridge between the canyons of inactivity – I would like to share ten tips that have helped me personally and continue to do so.[354]

Ten tips for walking the fine line of stoic action

- The climate crisis requires us to have a warm heart at least as much as a cool head. We must allow ourselves to be touched emotionally in order to move from the head through the heart into action. Recognizing, accepting, and processing our own **climate emotions** – such as sadness, anger, helplessness, or fear – is not only necessary but also deeply beneficial. By acknowledging these feelings, we can share, reflect on, and work through them with climate-conscious and supportive confidants, whether family members, partners, friends, or professional therapists.[355]

- We should **not demand perfection** when taking action. If a handprint idea is good enough for now and safe enough to try – go for it! A score of C+ is enough; only inaction is insufficient. More often than not, we will stagger and stumble along the fine line of stoic action rather than stride forward gracefully. But we should stumble as elegantly as possible. That means getting up more times than we fall. And then three steps forward, two steps back, and so on, until you reach your next milestone. And most importantly, we should celebrate the milestones along the way.

- We should also trust that **every action has an impact**. For every force, there is an effect. This holds true even when the ultimate effects of our actions lie beyond our direct field of vision – such as when we expand our collective handprint. Just like ripples from a pebble thrown into a lake, the waves of our actions will reach far shores, even if the fog of complexity obscures our view.

- When someone asks me 'Do you still have any hope that we can contain the climate crisis in time?', I answer: 'My head says no, my heart says yes.'[356] Hope is not about figures, data and facts, but about what you do yourself. **Hope comes from action**, not the other way around. This kind of hope isn't about analysing the probability of a desired future. Instead, it is rooted in a deep aspiration to reduce the suffering of fellow humans and living beings. It is action fueled and sustained by loving-kindness, compassion, joy, and equanimity. And you shouldn't let anyone talk you out of this idealistic commitment – especially not those who won't live to see the second half of this century themselves. (As Dutch historian and bestselling author Rutger Bregman aptly put it: 'Cynicism is another word for laziness.')[357]

- But be careful: fighting for a habitable planet should be neither a sprint nor a marathon – you'll likely burn out. As we move into the coming decades, facing escalating environmental (and consequently economic, political, and social) turmoil, our mindset should be more like hiking up a steep mountain together with a group. Step by step, neither rushing nor lingering, but pausing now and then to appreciate the view, our breath and how far we've come on our journey. **Resilience and self-care** will become the essential competencies in this century. The foundations of mental health include sufficient sleep, exercise, access to nature and a healthy diet. For me personally, regularly switching off, i.e. temporarily withdrawing completely from all climate issues, news and social media platforms, also helps me to recover.

- Strong and reliable social relationships, communities, and networks – whether within families, neighborhoods, friendships, clubs, or beyond – are so vital that they deserve a point of their own. First, personal relationships are a key pillar of mental health. And second, we will only be able to navigate the increasingly severe impacts of climate change within **strong, resilient communities.** The overwhelming local cooperation seen in most areas affected by climate disasters provides a glimpse of just how crucial this will become.

- This book has focused almost exclusively on reducing greenhouse gases. However, **adaptation** is just as crucial, as the impacts of climate change and extreme weather events will inevitably intensify. Moreover, to achieve true sustainability, we must address the intertwined social and environmental crises together and build a regenerative economy, ensuring we live **within both planetary and social boundaries.**[358] The British economist Kate Raworth captures this vividly in her Doughnut Economy model.[359]

- Ultimately, as industrial societies shaped by Western thought and monotheistic religions, we must dissolve the long-held dichotomy between nature and ('superior') humanity. We must rediscover 'Deep Ecology', our spiritual reverence for life, and remember that we are an integral part of our living environment – our Mother Earth, 'Gaia'.[360] Vietnamese peace activist, Buddhist monk, and Zen master Thich Nhat Hanh called this necessary realisation **'Interbeing'.**[361] With this understanding, qualities like compassion, respect, and humility naturally emerge.

- The understanding and mindset of *Interbeing* leads to a deep **sense of responsibility** for younger and future generations. After all, we have not inherited our place within the family of all things and creatures from our parents so much as we have borrowed it from our children.[362] When children ask me in the future, 'Gabriel, what did you do about the climate crisis in the 2020s?', I want to be able to answer honestly: 'I helped and expanded my handprint – without losing myself or my ability to feel joy. Everything else I had to learn to let go.'

- We should inwardly **let go** of everything that is beyond our control – everything outside our sphere of influence. Personally, my daily meditation practice helps me to do this: to be still, to watch my breathing, and to let go of everything else without judgement – again and again, as a simple yet not easy exercise. 'Keep calm, and happy, and carry on' – despite the climate crisis.

With these ten points in mind, I would like to conclude this book by giving the floor one last time to the head of the United Nations, Antonio Guterres: 'Humanity has a choice: cooperate or perish ... The global climate fight will be won or lost in this crucial decade – on our watch. One thing is certain: those that give up are sure to lose. So let's fight together – and let's win.'[363]

Then, in 2050, we can look back with joy and gratitude at having made a positive contribution to humanity's biggest and most important challenge, a kind of global wonder of the world. So, it's now or never! All hands on deck! Let's go!

Thanks

Only one name appears on the cover of this book, but no book is ever the work of just one person. As an author, I collaborated with many people to bring this book into the world. And in terms of content, I stand on the shoulders of giants, as the saying goes.

I cannot name everyone to whom I am grateful, nor can I personally thank all those whose snippets of conversation, advice, and insights have found their way into these pages. Yet, I feel a profound sense of gratitude toward them all. I hope that, in some way, my appreciation reaches them – perhaps through the same unexpected and indirect paths by which the influence of climate handprints leave their marks.

At the same time, I know for certain that a few people played a central role in the creation of this book: I would like to thank the teams at Edition Michael Fischer and at Canbury Press for their harmonious and constructive collaboration. In particular, I would like to thank Franziska Beyer, Iris Rinser and Martin Hickman for their close support and thoughtful engagement with my texts. The writing coach Janina Lücke, along with book agents Annette Brüggemann, Karoline Kuhn, and Annette Friese, provided invaluable guidance – especially in the early stages of this project, e.g. helping me navigate the process of finding a publisher.

The content of my book is deeply rooted in the Germanwatch handprint concept. I am profoundly grateful to all the handprint pioneers at Germanwatch, especially Marie Heitfeld, Alexander Reif, Carina Spieß, Daniela Baum, and Stefan Rostock. Their time and feedback greatly clarified and improved my explanations of the handprint approach.

Beyond Germanwatch, I was fortunate to have dedicated proofreaders supporting me throughout the writing process. My heartfelt thanks go to Janna Hoppmann, Daniel Obst and Jana Petersen, who spent countless hours reading and commenting on my ideas and drafts. This circle of support also includes Lea Dohm, Theresa Krüger, Sebastian Schels, Janine Steeger, Fridtjof Detzner and Eckart von Hirschhausen. Not only are all of them true Handprint role models, but each of them has personally supported me in different ways at various stages. I am deeply grateful to them all.

I also owe special thanks to Johannes Bosse and especially Jan Jöres, whose research efforts helped make this book possible. And what would a nonfiction book be without compelling, thought-provoking, or smile-inducing graphics? For that creative and artistic achievement, I, and you, have Katharina Steiner to thank.

This book would not exist without Climaware, so my gratitude extends to everyone who has contributed to its success – especially my podcast producers and friends Lukas Schreiber and Tim Kleikamp, as well as Valerie Helbich-Poschacher. Of course, I also thank all the interview guests who took the time to share their insights on the show. The same applies to the podcast Über Klima sprechen, based on the handbook of the same name by Christopher Schrader and the team at Klimafakten. A big thank you to them as well for their contributions and inspiration on climate communication.

Writing this book was filled with moments of joy and creative flow, but it also cost me blood, sweat, and tears – literally. Corina, I thank you for being my light in dark times, for your unwavering emotional support, for the countless precious hours of reading and conversation, and, most of all, for your love. This also applies to my close friends, my siblings, parents and grandparents.

Finally, I want to acknowledge all the courageous and kindred spirits out there who, with their minds, hearts, and hands, are working to protect our precious planet. You are the ones who actively make hope happen. And thank you, dear readers, for the time and attention you have given to me and my text.

Appendix

My personal list of recommendations for further climate education

Climate science & data
- IPCC Synthesis Report (2023) – The latest summary of global climate science
- Emissions Gap Report (2024), UNEP – Tracks the policy gaps to our climate goals
- Climate Action Tracker (climateactiontracker.org)

Climate news
- CarbonBrief (website & newsletter)
- Inside Climate News (website & newsletter)
- Environment section of *The Guardian*

Climate solutions
- Project Drawdown (drawdown.org)
- The Climate Game, Financial Times (online game: ig.ft.com/climate-game/)
- En-ROADS Simulator (climateinteractive.org/en-roads/)

Non-fiction books
- *The Future We Choose* (2021), Christiana Figueres & Tom Rivett-Carnac
- *Zen and the Art of Saving the Planet* (2021), Thich Nhat Hanh
- *Less is More* (2020), Jason Hickel
- *Citizens* (2022), Jon Alexander
- *Active Hope* (2012, revised 2022), Joanna Macy & Chris Johnstone
- *Land Sickness* (2023), Nikolaj Schultz

Films & documentaries
- Before the Flood (2016) – Leonardo DiCaprio explores the climate issue
- Our Planet (2019) – David Attenbourough shows the beauty and fragility of our planet
- Don't Look Up! (2022) – A satirical film about the societal denial of the climate crisis
- 2040 (2019) – Damon Gameau's positive documentary of climate solutions

Podcasts
- Outrage + Optimism: The Climate Podcast, Christiana Figueres, Tom Rivett-Carnac & Paul Dickinson
- Plum Village's podcast The Way Out Is In, Brother Phap Huu & Jo Confino
- English episodes of Climaware, Gabriel Baunach

More about the handprint
- Germanwatch (germanwatch.org/en/handprint)
- Global Handprint Network (handprint.in/)
- Follow Life-Cycle Assessment (LCA) expert Gregory Norris
-

Other helpful organisations & resources
- Climate Outreach (climateoutreach.org)
- Inner Development Goals (innerdevelopmentgoals.org)
- Work That Reconnects (workthatreconnects.org)
- Zen and the Art of Saving the Planet, online course by Plum Village

Endnotes

1   https://www.un.org/sg/en/content/sg/speeches/2022-11-07/secretary-generals-remarks-high-level-opening-of-cop27, accessed 04.02.2025

2   https://www.nature.com/articles/s41586-023-06970-0 & https://www.science.org/doi/10.1126/sciadv.adk1189, accessed 31.01.2025

3   I still have the film sequences from *The Day after Tomorrow* in my mind's eye. The 2004 blockbuster impressively depicted the tipping of the Gulf Stream caused by global warming (albeit somewhat exaggerated in Hollywood style).

4   "Exceeding 1.5°C global warming could trigger multiple climate tipping points", D. I. A. McKay, et al., 2022, Link: https://www.science.org/doi/10.1126/science.abn7950, accessed 24.04.2023

5   See https://www.theguardian.com/environment/2023/feb/14/rising-seas-threaten-mass-exodus-on-a-biblical-scale-un-chief-warns, accessed 24.04.2023

6   Engels, Anita; Jochem Marotzke; Eduardo Gonçalves Gresse; Andrés López-Rivera; Anna Pagnone; Jan Wilkens (eds.); 2023. Hamburg Climate Futures Outlook 2023. The plausibility of a 1.5°C limit to global warming-Social drivers and physical processes. Cluster of Excellence Climate, Climatic Change, and Society (CLICCS). Hamburg, Germany

7   United Nations Environment Program (2022). Emissions Gap Report 2022: The Closing Window - Climate crisis calls for rapid transformation of societies. Nairobi. https://www.unep.org/emissions-gap-report-2022 & https://climateactiontracker.org/global/temperatures/, accessed 24.04.2023

8   See for example https://climateactiontracker.org/countries/germany/ & https://climateactiontracker.org/countries/switzerland/, accessed 24.04.2023

9   https://www.ipcc.ch/report/sixth-assessment-report-cycle/, accessed 04.02.2025

10   https://www.un.org/sg/en/content/sg/speeches/2022-11-07/secretary-generals-remarks-high-level-opening-of-cop27, accessed 04.02.2025

11   https://www.undp.org/press-releases/80-percent-people-globally-want-stronger-climate-action-governments-according-un-development-programme-survey, accessed 31.01.2025

12   United Nations Environment Program (2022). Emissions Gap Report 2022: The Closing Window - Climate crisis calls for rapid transformation of societies. Nairobi. https://www.unep.org/emissions-gap-report-2022, accessed 24.04.2023

13   https://www.gov.uk/government/statistics/desnz-public-atti-tudes-tracker-spring-2024/desnz-public-attitudes-tracker-net-zero-and-climate-change-spring-2024-uk?utm_source=chatgpt.com & https://www.ons.gov.uk/peoplepopulationandcommunity/wellbeing/articles/worries-aboutclimatechangegreatbritain/septembertooctober2022 & https://www.nature.com/articles/s41558-024-01925-3, accessed 07.01.2025

14   https://www.nature.com/articles/s41558-024-01925-3 & https://cli-matecommunication.yale.edu/publications/climate-change-in-the-ameri-can-mind-april-2022/ & https://www.pewresearch.org/fact-tank/2019/04/18/a-look-at-how-people-around-the-world-view-climate-change/ & https://www.oecd.org/climate-change/international-attitudes-toward-climate-pol-icies/, accessed 24.04.2023

15   See article at https://www.klimareporter.de/gesellschaft/verzer-rte-wahrnehmung, accessed 24.04.2023, sources: "Americans experience a false social reality by underestimating popular climate policy support by nearly half", Parkman, et al, 2022, Nature, Link: https://www.nature.com/articles/s41467-022-32412-y, accessed 24.04.2023 & https://projekte.uni-er-furt.de/pace/_files/PACE_W07-09.pdf#page=24, accessed 24.04.2023

16   For a scientific description of the "sedative pill" and the "epis-temic fit", see "Why we should Empty Pandora's Box to Create a Sustain-able Future: Hope, Sustainability and Ist Implications for Educations", Gr-und, et. al., 2019, Link: https://www.mdpi.com/2071-1050/11/3/893, accessed 24.04.2023

17   "Young People's Voices on Climate Anxiety, Government Betrayal and Moral Injury: A Global Phenomenon", Marks, et al, 2021, Nature, Link https://papers.ssrn.com/sol3/papers.cfm?abstract_id=3918955, accessed 24.04.2023, accessed 24.04.2023

18   https://www.ons.gov.uk/economy/environmentalaccounts/arti-cles/climatechangeinsightsuk/august2023, accessed 07.01.2025

19   Creutzig, F., J. Roy, P. Devine-Wright, J. Díaz-José, F.W. Geels, A. Grubler, N. Maïzi, E. Masanet, Y. Mulugetta, C.D. Onyige, P.E. Perkins, A. Sanches-Pereira, E.U. Weber, 2022: Demand, services and social aspects of mitigation. In IPCC, 2022: Climate Change 2022: Mitigation of Climate Change. Contribution of Working Group III to the Sixth Assessment Report of the Intergovernmental Panel on Climate Change [P.R. Shukla, J. Skea, R. Slade, A. Al Khourdajie, R. van Diemen, D. McCollum, M. Pathak, S. Some, P. Vyas, R. Fradera, M. Belkacemi, A. Hasija, G. Lisboa, S. Luz, J. Malley, (eds.)]. Cambridge University Press, Cambridge, UK and New York, NY, USA. doi: 10.1017/9781009157926.007

20   For the whole paragraph: https://handprint.in & https://www.ceeindia.org/sdg-handprint-lab & https://unevoc.unesco.org/home/4th+In-ternational+Conference+on+Environmental+Education+in+Ahmed-

abad,+India & background paper "Wandel mit Hand und Fuß", Marie
Heitfeld and Alexander Reif, Germanwatch, 2015, Link: https://www.
germanwatch.org/de/12040 & "Shaping Transformation", Marie Heitfeld
and Alexander Reif, Germanwatch, 2020, Link: https://www.germanwatch.
org/de/19607, accessed 24.04.2023

21   Some people use the term "CO2 handprint" instead of "climate
handprint" (and I have also come across the term "climate shadow"). In
principle, it means the same thing, but CO2 is not the only greenhouse
gas. Furthermore, the Germanwatch handprint concept is not only related
to emissions, but also to a structurally changing commitment to more
ecologically and socially sustainable framework conditions.

22   Norris, GA, 2013. The New Requirement for Social Leadership:
Healing.
in S. Groschl (ed.).Uncertainty, Diversity and the Common Good: Changing Norms and New Leadership Paradigms. London: Gower Publishing,
link: https://shine.mit.edu/sites/default/files/Norris%202013%20An%20In-
troduction%20to%20Handprints%20and%20Handprinting_2.pdf, see also
his TEDx talk: https://www.youtube.com/watch?v=qvvd9qvs9Es, accessed
04.02.2025

23   https://www.atmosfair.de/de/kompensieren/flug/ & "So viel
CO_2 stoßen Autos - Geschätzte durchschnittliche CO2 -Emissionen von
PKW in Deutschland 2022 (in kg CO2 /Jahr)", M. Janson, Link: https://
de.statista.com/infografik/25742/durchschnittliche-co2-emission-von-pkw-
in-deutschland-im-jahr-2020/, accessed 24.04.2023

24   "Consumer behavior and climate change: consumers need
considerable assistance", John Thogersen, 2021, Current Opinion in Behavioral Sciences, Link: https://www.sciencedirect.com/science/article/pii/
S2352154621000309, accessed 24.04.2023

25   Assumptions & calculation: The conversions mean that 300
meat dishes with around 1 kg CO2 eq. each (i.e. a total of 60,000 kg CO2
eq. per year) are avoided on 200 days a year.

26   Assumptions & calculation: 250,000 kWh with a CO2 factor for
2023 of 0.225 kg CO2 per kWh in the British electricity mix are replaced
by (almost) carbon-neutral solar power. Source: UK Department for Business, Energy & Industrial Strategy (BEIS)

27   With a CO2 factor for the British electricity mix for 2023 of 0.225
kg CO2/kWh, source: UK Department for Business, Energy & Industrial
Strategy (BEIS)

28   Chassot, Sylviane; Wüstenhagen, Rolf; Fahr, Nicole & Graf,
Peter: "Wenn das grüne Produkt zum Standard wird : Wie ein Energiever-
sorger den Kunden die Verhaltensänderung einfach macht." In: Organiza-
tional Development (2013), 3, pp. 80-87. link: https://www.researchgate.net/

publication/268213904_Wenn_das_grune_Produkt_zum_Standard_wird_Wie_ein_Energieversorger_seinen_Kunden_die_Verhaltensanderung_einfach_macht, accessed 25.04.2023

29   "Domestic uptake of green energy promoted by opt-out tariffs", Ebeling, et al., 2015, Link: https://www.nature.com/articles/nclimate2681, accessed 25.04.2023

30   An environmental psychology study in 2018, for example, showed that community engagement in sustainability (an increase in handprint) makes personal energy saving (a reduction in footprint) more likely. Source: "Can community energy initiatives motivate sustainable energy behaviors? The role of initiative involvement and personal pro-environmental motivation", Sloot, et al., 2018, Journal of Environmental Psychology Link: https://www.sciencedirect.com/science/article/abs/pii/S027249441830330X, accessed 25.04.2023

31   Interview with Prof. Dr. Stefan Rahmstorf in the Climaware Podcast: https://open.spotify.com/episode/4NeZxInOnpjrTI8sooSEhq, accessed 25.04.2023

32   https://sea.mashable.com/science/11514/the-carbon-footprint-sham, accessed 25.01.2025

33   The story is described in a New York Times Magazine article from February 17, 2008 by William Safire: https://www.nytimes.com/2008/02/17/magazine/17wwln-safire-t.html, accessed 04/25/2023

34   "Ecological footprints and appropriated carrying capacity: what urban economics leaves out", Rees, 1992, Link: https://journals.sagepub.com/doi/10.1177/095624789200400212, Zugriff 25.04.2023

35   See https://www.footprintnetwork.org/, accessed 25.04.2023

36   Tom Rawls in an interview for the article "Scientists count carbon in global-warming fight" by Eric Sorensen for the newspaper "The Seattle Times" on 13.11.2000, Link: https://archive.seattletimes.com/archive/?date=20001113&slug=4052870, Zugriff 24.04.2023

37   "Assessing ExxonMobil's global warming projections", Supran et al, 2023, Science, Link: https://www.science.org/doi/10.1126/science.abk0063, accessed 24.04.2023

38   https://www.theguardian.com/environment/2021/nov/18/the-forgotten-oil-ads-that-told-us-climate-change-was-nothing, accessed 24.04.2023

39   see for example the books "The Merchants of Doubt", Naomi Oreskes and Erik M. Conway, 2010 & "Die Klimaschmutzlobby", Susanne Götze and Annika Joeres, Piper, 2022

40   Own translation for "American Marketing Association". Source: https://foreignpolicy.com/2010/05/03/back-to-petroleum/, accessed 24.04.2023

41   https://www.nytimes.com/2006/08/14/opinion/14kenney.html, accessed 24.04.2023

42   https://www.theguardian.com/environment/2015/apr/16/bp-dropped-green-energy-projects-worth-billions-to-focus-on-fossil-fuels, accessed on 24.04.2023

43   https://en.wikipedia.org/wiki/BP#Violations_and_accidents, accessed on 24.04.2023

44   4 million barrels per day in 2005: BP Annual Report 2005, Link: https://ddd.uab.cat/pub/infanu/43618/iaBPa2005ieng2.pdf & 3.8 million barrels per day in 2019: BP Annual Report 2019, Link: https://www.bp.com/content/dam/bp/country-sites/nl-nl/netherlands/home/documents/corporate-and-finance/bp-annual-report-and-form-20f-2019.pdf & https://www.theguardian.com/business/2023/feb/07/bp-profits-windfall-tax-gas-prices-ukraine-war, accessed 25.04.2023

45   USD 200 million: https://www.prwatch.org/news/2010/05/9038/bps-beyond-petroleum-campaign-losing-its-sheen & https://foreignpolicy.com/2010/05/03/back-to-petroleum/, accessed 17.02.2022 & USD 370 million: https://www.adweek.com/brand-marketing/bp-coloring-public-opinion-91662/, accessed 25.04.2023

46   https://www.adweek.com/brand-marketing/bp-coloring-public-opinion-91662/, accessed 25.04.2023

47   https://www.theguardian.com/environment/2021/nov/18/the-forgotten-oil-ads-that-told-us-climate-change-was-nothing, accessed 25.04.2023

48   BP alone is responsible for more than 2 percent of all anthropogenic greenhouse gas emissions from 1751 to 2018. Source: https://climate-accountability.org/?page_id=18, accessed 30.09.22 & "The Carbon Majors Database - CDP Carbon Majors Report 2017" https://www.cdp.net/en/articles/media/new-report-shows-just-100-companies-are-source-of-over-70-of-emissions, accessed 25.04.2023

49   Average Scope 3 emissions for the years 2019-2021, source: https://www.globaldata.com/data-insights/macroeconomic/bp-annual-ghg-emissions/, accessed 25.04.2023

50   To calculate: 10 tons (average individual emissions in Germany per year) to 376,000,000 tons (average annual Scope 3 emissions from BP in the period 2019-2021) is a ratio of 1:37,000,000. A human footprint of approx. 0.025 m2 therefore means - multiplied by 37,000,000 - a BP footprint of approx. 940,000 m2 . According to the DFB, the standard size of a soccer pitch is 7,140 m2 . The Cheops pyramid in Egypt has a base area of approx. 52,900 m2 .

51   Source for the entire section: https://www.adweek.com/brand-marketing/bp-coloring-public-opinion-91662/, accessed 25.04.2023

52   Searches with the Google Books NGram Viewer with the search terms "carbon footprint" in the corpus "English (2019)" and "CO2 footprint" in the corpus "German (2019)" & Searches with the search term "carbon footprint" (compared to the search term "climate action") in the region "worldwide" and "United States" (2004-present; all categories; web search) at Google Trends, accessed 21.02.2022

53   William Safire, "Footprint", The New York Times Magazine, 17.02.2008, https://www.nytimes.com/2008/02/17/magazine/17wwln-safire-t.html, translated with DeepL.com/translator, accessed 25.04.2023

54   "Conceptualizing Guilt in the Consumer Decision-making Process", Burnett, et al., 1994, Link: https://www.emerald.com/insight/content/doi/10.1108/07363769410065454/full/html, accessed 25.04.2023 & "Does impact of campaign and consumer guilt help in exploring the role of national identity and purchase decisions of consumers?", Malhotra, et al., 2022, Link: https://www.sciencedirect.com/science/article/abs/pii/S0969698921004057, accessed 25.04.2023

55   https://www.theregreview.org/2018/11/05/braga-cook-guns-do-kill-people/, accessed 25.04.2023

56   The advertisement can still be seen, for example, on the YouTube channel coffeekid99: https://www.youtube.com/watch?v=-j7OHG7tHrNM, accessed 25.04.2023

57   https://www.chicagotribune.com/opinion/commentary/ct-perspec-indian-crying-environment-ads-pollution-1123-20171113-story.html, accessed 25.04.2023

58   "Rhetoric and frame analysis of ExxonMobil's climate change communications", G. Supran and N. Oreskes, 2021, Link: https://www.sciencedirect.com/science/article/pii/S2590332221002335, accessed 25.04.2023

59   Current examples e.g. here: https://grist.org/beacon/shells-false-carbon-offset-promises/, accessed 25.04.2023

60   Tweet from 22.10.2019 via the official account @bp_plc, sources: https://twitter.com/bp_plc/status/1186645440621531136?lang=en & https://www.knowyourcarbonfootprint.com, accessed 25.04.2023

61   https://www.theguardian.com/environment/2022/sep/08/oil-and-gas-firms-green-investments-fail-to-match-promise-of-adverts-study, accessed 25.04.23

62   See for example https://nymag.com/intelligencer/2020/03/shell-climate-change.html & https://medium.com/@sami.grover/in-defense-of-eco-hypocrisy-b71fb86f2b2f, accessed 25.04.2023

63   https://www.theguardian.com/environment/ng-interactive/2022/may/11/fossil-fuel-carbon-bombs-climate-breakdown-oil-gas, accessed 25.04.2023

64   https://oilchange.org/borc/, access 07.01.2025

65   "Nudging out support for a carbon tax", Hagman, et al, 2019, Nature Climate Change, Link: https://www.nature.com/articles/s41558-019-0474-0 & "Household behavior crowds out support for climate change policy when sufficient progress is perceived", Werfel, 2017, Nature, Link: https://www.nature.com/articles/nclimate3316 & "Why hate carbon taxes? Machine learning evidence on the roles of personal responsibility, trust, revenue recycling, and other factors across 23 European countries", Levi, 2021, Energy Research & Social Science, Link: https://www.sciencedirect.com/science/article/abs/pii/S2214629620304588?via%3Dihub, accessed 25.04.2023

66   "How do climate change skeptics engage with opposing views? Understanding mechanisms of social identity and cognitive dissonance in an online forum", Lisa Oswald and Jonathan Bright, 2021, Link: https://www.researchgate.net/publication/349309444_How_do_climate_change_skeptics_engage_with_opposing_views_Understanding_mechanisms_of_social_identity_and_cognitive_dissonance_in_an_online_forum, accessed 25.04.2023

67   https://klimakommunikation.klimafakten.de/vor-denken/kapitel-1-mach-dir-klar-was-bisher-schiefgelaufen-ist/, accessed 25.04.2023

68   "Interested, indifferent or active information avoiders of carbon labels: Cognitive dissonance and ascription of responsibility as motivating factors", A. K. Edenbrandta, et al, Food Policy 101, 2021

69   See for example "Scepticism and uncertainty about climate change: Dimensions, determinants and change over time", Lorraine Whitmarsh, Global Environmental Change 21, p. 690-700, 2011

70   See for example handbook "Über Klima sprechen" by Christopher Schrader & klimafakten.de, Oekom-Verlag, 2022 & "Environmental Knowledge, Attitudes, and Behavior in Dutch Secondary Education", Kuhlemeier, et al., 1999, Link: https://www.researchgate.net/publication/240538886_Environmental_Knowledge_Attitudes_and_Behavior_in_Dutch_Secondary_Education, accessed 25.04.2023

71   Introduction to the handbook "Über Klima sprechen" by Christopher Schrader & klimafakten.de published by Oekom-Verlag, 2022, Link: https://klimakommunikation.klimafakten.de/einleitung/, accessed 25.04.2023

72   "Environmental awareness in Germany 2020, results of a representative population survey, BMUV & UBA, 2022, Link: https://www.bmuv.de/fileadmin/Daten_BMU/Pools/Broschueren/umweltbewusstsein_2020_bf.pdf, accessed 25.04.2023

73   And if I may dream a little: perhaps the handprint concept will ultimately convince even a few of those who are still unwilling, because it

is not so much about their individual guilt or their supposed renunciation, but a bypass.

74   "The climate mitigation gap: education and government recommendations miss the most effective individual actions", Wynes, et al, 2017, Link: https://iopscience.iop.org/article/10.1088/1748-9326/aa7541 & https://www.klimafakten.de/kommunikation/klimakommunikation-auf-massnahmen-fokussieren-auf-wenige-und-die-wirksamsten-fuenf & https://www.de.kearney.com/social-impact/article/-/insights/kein-plan-vom-klimaschutz-, accessed 25.04.2023

75   "Climate Change + Consumer Behavior. A Global Advisor Survey", Ipsos, October 2021, Link: https://www.ipsos.com/en/climate-change-consumer-behaviour-2021, accessed 25.04.2023

76   https://www.klimafakten.de/kommunikation/klimakommunikation-auf-massnahmen-fokussieren-auf-wenige-und-die-wirksamsten-fuenf & "Consumer behavior and climate change: consumers need considerable assistance", Thogersen, 2021, Current Opinion in Behavioral Sciences, Link: https://www.sciencedirect.com/science/article/pii/S2352154621000309 & "Interested, indifferent or active information avoiders of carbon labels: Cognitive dissonance and ascription of responsibility as motivating factors", A. K. Edenbrandta, et al, Food Policy 101, 2021, accessed 25.04.2023.

77   "Climate Action - Psychology of the climate crisis. Obstacles to action and opportunities for action", Lea Dohm, Felix Peter, Katharina van Bronswijk (eds.), Psychosozial-Verlag, 2021

78   See for example: "A Meta-Analytic Review of Moral Licensing", I. Blanken, et al., 2015, Link: https://journals.sagepub.com/doi/pdf/10.1177/0146167215572134?casa_token=A25xER-rhV8AAAAA:qjRQPEZIfPar-TINYscbFCBmMsoPxVwUweOxo4yqv9PIoKdWjieVK68T-Uk5z-PRuYfd8espKnTw & "Moral Self-Licensing: When Being Good Frees Us to Be Bad", Anna C. Merritt, et al, 2010, Link: https://compass.onlinelibrary.wiley.com/doi/10.1111/j.1751-9004.2010.00263.x, Zugriff 25.04.2023

79   See for example "I did my bit! The impact of electric vehicle adoption on compensatory beliefs and norms in Norway", Nayum, 2022, Energy Research & Social Science, Link: https://www.sciencedirect.com/science/article/pii/S2214629622000482, accessed 25.04.2023

80   "For better or for worse? Empirical evidence of moral licensing in a behavioral energy conservation campaign", Tiefenbeck, et al, 2013, Energy Policy, Link: https://www.sciencedirect.com/science/article/pii/S0301421513000281?via%3Dihub, accessed 25.04.2023

81   "The rebound effect. On the undesirable consequences of desirable energy efficiency", Tilman Santarius, Wuppertal Institute for Climate, Environment and Energy GmbH, Impulse zur WachstumsWende,

March 2012, Link: http://www.santarius.de/wp-content/uploads/2012/03/
Der-Rebound-Effekt-2012.pdf & "Empirical estimates of the direct rebound effect: A review", S. Sorrell, et al, 2009, Link: https://www.science-
direct.com/science/article/pii/S0301421508007131?via%3Dihub & "The
rebound effect in road transport: A meta-analysis of empirical studies", A.
Dimitropoulos, et al., 2018, Link: https://www.sciencedirect.com/science/
article/pii/S0140988318302718, accessed 25.04.2023

82   "Rebound effect of efficiency improvement in passenger cars on
gasoline consumption in Canada", Saeed Moshiri and Kamil Aliyev, Ecological Economics, Vol. 131, January 2017, Link: https://www.sciencedirect.
com/science/article/pii/S0921800915303438, accessed 25.04.2023

83   This case study appears in a scientific study on rebound effects
from 2013: "Turning lights into flights: Estimating direct and indirect rebound effects for UK households", Chitnis, et al, 2013, Energy Policy, Link:
https://www.sciencedirect.com/science/article/abs/pii/S0301421512010531 &
"Prospects for radical emissions reduction through behavior and lifestyle
change", Capstick, 2015, Carbon Management, Link: https://www.tandfon-
line.com/doi/full/10.1080/17583004.2015.1020011, accessed 25.04.2023

84   CO2 emissions of SUVs: https://www.greenpeace.de/pub-
likationen/s03141_es_gp_report_suv_12_2020.pdf & CO2 eq. savings of
regional food: "Ecological footprints of food and dishes in Germany",
Wagner, et al., ifeu study, 2020, Link: https://www.ifeu.de/fileadmin/
uploads/Reinhardt-Gaertner-Wagner-2020-Oekologische-Fu%c3%9fab-
druecke-von-Lebensmitteln-und-Gerichten-in-Deutschland-ifeu-2020.
pdf, accessed 25.04.2023

85   Recordings can be seen on YouTube, for example: https://www.
youtube.com/watch?v=KE5YwN4NW50, accessed 25.04.2023

86   Latane, B., & Darley, J. M. (1968). Group inhibition of bystander
intervention in emergencies. Journal of Personality and Social Psychology,
10(3), 215-221, Link: https://psycnet.apa.org/record/1969-03938-001, Zugriff
25.04.2023

87   A meta-study from 2011 found four main causes for the bystand-
er effect: pluralistic ignorance, diffusion of responsibility, one's own (in)
ability assessment and observer anxiety. Source: Peter Fischer et al, "The
Bystander Effect. A Meta-Analytic Review on Bystander Intervention in
Dangerous and Non-Dangerous Emergencies". Psychological Bulletin,
Vol. 137, Issue 4, 2011

88   Latane, B., & Darley, J. M. (1968). Group inhibition of bystander
intervention in emergencies. Journal of Personality and Social Psychology,
10(3), 215-221, Link: https://psycnet.apa.org/record/1969-03938-001, Zugriff
25.04.2023

89   https://en.wikipedia.org/wiki/Beds_Are_Burning, accessed 25.04.2023

90   See for example: https://www.sciencedirect.com/topics/psychology/intention-behavior-gap, accessed 25.04.2023

91   https://open.spotify.com/show/28sR8OiOqoMMnGEzMJTXSt, accessed 25.04.2023

92   "The social shortfall and ecological overshoot of nations", A. L. Fanning, et al, Nature Sustainability, 2021, Link: https://www.nature.com/articles/s41893-021-00799-z, Zugriff 25.04.2023

93   The perceived cost, or effort, of climate-friendly behavior and the perceived climate effect strongly influence whether a person takes action against the climate crisis themselves or supports climate protection policy measures. See for example: "Addressing climate change: Determinants of consumers' willingness to act and to support policy measures", Christina Tobler, et al, Journal of Environmental Psychology, Vol. 32, Issue 3, p. 197-207, 2012, Link: https://www.sciencedirect.com/science/article/abs/pii/S0272494412000102?via%3Dihub & "Systematic Literature Review on Behavioral Barriers of Climate Change Mitigation in Households", G. Stankuniene, et al., Sustainability, 2020 & "Taking Responsibility for Sustainable Behavior - An Empirical Analysis from a Social Psychological Perspective", H. Vogelsang and C. Buchholz, IZNE Working Paper Series No. 19/2, 2018, Link: https://pub.h-brs.de/frontdoor/deliver/index/docId/4395/file/IZNE_WP_1902.pdf, accessed 25.04.2023

94   Diekmann, A., & Preisendörfer, P. (2017). Personal environmental behavior discrepancies between aspiration and reality. Cologne Journal of Sociology and Social Psychology, pp. 591-617

95   According to the IMF, fossil fuel subsidies amounted to 5.9 trillion US dollars worldwide in 2020, which corresponds to around 6.8% of global GDP. In Germany, environmentally harmful subsidies amounted to over 65 billion euros, primarily in the transport and energy sectors. Sources: "Still Not Getting Energy Prices Right: A Global and Country Update of Fossil Fuel Subsidies", I. W. H. Parry, et al., 2021, Link: https://www.imf.org/en/Publications/WP/Issues/2021/09/23/Still-Not-Getting-Energy-Prices-Right-A-Global-and-Country-Update-of-Fossil-Fuel-Subsidies-466004 & "Umweltschädliche Subventionen in Deutschland", A. Burger and W. Bretschneider, Umweltbundesamt, 2021, Link: https://www.umweltbundesamt.de/sites/default/files/medien/479/publikationen/texte_143-2021_umweltschaedliche_subventionen.pdf, accessed 25.04.2023

96   https://lexikon.stangl.eu/18288/soziale-normen, accessed 25.04.2023

97   https://www.klimafakten.de/meldung/warum-gutmenschen-bei-manchen-leuten-so-unbeliebt-sind-und-was-sie-dagegen-tun-koennen, accessed 25.04.2023

98   "Minority influence in climate change mitigation", Bolderdijk, et al., 2021, Current Opinion in Psychology, Link: https://www.sciencedirect.com/science/article/pii/S2352250X21000154, Translated with www.DeepL.com/Translator (free version), accessed 25.04.2023

99   "Talking about climate. Das Handbuch", Christopher Schrader and klimafakten.de, Chapter 2: Know yourself - and your weaknesses, 2021, Link: https://klimakommunikation.klimafakten.de/vor-denken/kapitel-2-kenne-dich-selbst-und-deine-schwaechen/, accessed 25.04.2023

100   https://open.spotify.com/episode/3YpndTks1Fu8mkpGEkxa-He?si=fdc363005d4644db, accessed 25.04.2023

101   "Why Do We Hate Hypocrites? Evidence for a Theory of False Signaling", Jordan, et al, 2017, Link: https://journals.sagepub.com/doi/full/10.1177/0956797616685771, Zugriff 25.04.2023

102   https://www.youtube.com/watch?v=iV8y7PFJYgc, accessed 25.04.2023

103   https://open.spotify.com/episode/1nJTLeXwgHCKH4Md82AlfX-?si=b5cG9hafTSeOvsYaXQ7h7w, accessed 25.04.2023

104   https://www.netflix.com/title/81252357, accessed 25.04.2023

105   https://www.dailymail.co.uk/tvshowbiz/article-10382825/Dont-look-Leonardo-DiCaprio-eco-hypocrite-110million-yacht.html, accessed 25.04.2023

106   Although to be fair, I have to mention that Leonardo DiCaprio is taking steps to reduce his enormous CO_2 footprint: In everyday life, he mostly drives his Prius hybrid car and is now increasingly using scheduled flights, in 2021 for example, to fly to the UN Climate Change Conference in Glasgow.

107   https://www.youtube.com/watch?v=xpyrefzvTpI, accessed 25.04.2023

108   https://time.com/4441219/leonardo-dicaprio-oscars-climate-change/, accessed 25.04.2023

109   https://taz.de/Thunbergs-Segelreise-in-die-USA/!5615733/, accessed 25.04.2023

110   "Climate change and moral judgment", Ezra M. Markowitz and Azim F. Shariff, Nature Climate Change, Vol. 2, 2012

111   Environmental journalist David Roberts gets to the heart of the matter in an article: https://www.vox.com/2016/3/2/11143310/leo-dicaprios-carbon-lifestyle, accessed 25.04.2023

112   Festinger, L. (1957). A theory of cognitive dissonance. Stanford University Press, Link: https://psycnet.apa.org/record/1993-97948-000

& "Cognitive Dissonance and Consumer Behavior: A Review of the Evidence", William H. Cummings and M. Venkatesan, 1976, Link: https://www.jstor.org/stable/3150746, accessed 25.04.2023

113   See for this entire section on dissonance reduction: "Dealing with dissonance: A review of cognitive dissonance reduction", A. Mc-Grath, 2017, Link: https://compass.onlinelibrary.wiley.com/doi/abs/10.1111/spc3.12362, accessed 25.04.2023

114   Festinger, L. (1957). A theory of cognitive dissonance. Stanford University Press, Link: https://psycnet.apa.org/record/1993-97948-000, Zugriff 25.04.2023

115   https://www.umweltbundesamt.de/sites/default/files/medien/479/publikationen/texte_20-2022_repraesentativumfrage_zum_umweltbewusstsein_und_umweltverhalten_im_jahr_2020.pdf, accessed 25.04.2023

116   https://www.droemer-knaur.de/buch/petra-pinzler-guenther-wessel-vier-fuers-klima-9783426302736, accessed 25.04.2023

117   IPCC Assessment Report 6, 2022, Chapter 5: Creutzig, F., J. Roy, P. Devine-Wright, J. Díaz-José, F.W. Geels, A. Grubler, N. Maïzi, E. Masanet, Y. Mulugetta, C.D. Onyige, P.E. Perkins, A. Sanches-Pereira, E.U. Weber, 2022: Demand, services and social aspects of mitigation. In IPCC, 2022: Climate Change 2022: Mitigation of Climate Change. Contribution of Working Group III to the Sixth Assessment Report of the Intergovernmental Panel on Climate Change [P.R. Shukla, J. Skea, R. Slade, A. Al Khourdajie, R. van Diemen, D. McCollum, M. Pathak, S. Some, P. Vyas, R. Fradera, M. Belkacemi, A. Hasija, G. Lisboa, S. Luz, J. Malley, (eds.)]. Cambridge University Press, Cambridge, UK and New York, NY, USA. doi: 10.1017/9781009157926.007 & "The potential of behavioral change for climate change mitigation: a case study for the European Union", Dirk-Jan van de Ven, et al., 2017, Link: https://link.springer.com/article/10.1007/s11027-017-9763-y & " Prospects for radical emissions reduction through behavior and lifestyle change", Capstick, et al., 2015, Link: https://www.tandfonline.com/doi/full/10.1080/17583004.2015.1020011, accessed 26.04.2023 & "Household actions can provide a behavioral wedge to rapidly reduce US carbon emissions", Dietz, et al., 2009, PNAS, Link: https://www.pnas.org/doi/full/10.1073/pnas.0908738106, accessed 26.04.2023 & "A review of intervention studies aimed at household energy conservation", Abrahamse, 2009, Journal of Environmental Psychology, Link: https://www.sciencedirect.com/science/article/abs/pii/S027249440500054X?via%-3Dihub, accessed 26.04.2023 & The effectiveness of soft transport policy measures: A critical assessment and meta-analysis of empirical evidence, Möser, 2008, Journal of Environmental Psychology, Link: https://www.sciencedirect.com/science/article/abs/pii/S0272494407000722?via%-3Dihub, accessed 26.04.2023 & "Promoting walking and cycling as an

alternative to using cars: systematic review", Ogilvie, et al, 2004, Link: https://www.bmj.com/content/329/7469/763, accessed 26.04.2023 & https://assets.publishing.service.gov.uk/government/uploads/system/uploads/attachment_data/file/69797/6921-what-works-in-changing-energyusing-behaviours-in-.pdf, accessed 26.04.2023 & "Exploring the implications of lifestyle change in 2 °C mitigation scenarios using the IMAGE integrated assessment model", Sluisveld, et al, 2016, Technological Forecasting and Social Change, Link: https://www.sciencedirect.com/science/article/abs/pii/S0040162515002607?via%3Dihub, accessed 26.04.2023 & "Real-time feedback promotes energy conservation in the absence of volunteer selection bias and monetary incentives", Tiefenbeck, et al, 2018, Nature Energy, Link: https://www.nature.com/articles/s41560-018-0282-1, accessed 26.04.2023 & "Motivating energy conservation in the workplace: An evaluation of the use of group-level feedback and peer education", Carrico, et al, 2011, Elsevier Journal of Environmental Psychology, Link: https://www.sciencedirect.com/science/article/abs/pii/S0272494410001015?via%3Dihub, accessed 26.04.2023 & https://drawdown.org/news/insights/the-powerful-role-of-household-actions-in-solving-climate-change, accessed 26.04.2023 & "2018 Climate Change Need Behavior Change", Rare, Link: https://rare.org/report/climate-change-needs-behavior-change/, accessed 26.04.2023 & "The Power of People - Climate Action and the Role of Citizens", The JUMP, 2022, Link: https://takethejump.org/power-of-people, accessed 26.04.2023

118   conservation in the workplace: An evaluation of the use of group-level feedback and peer education", Carrico, et al, 2011, Elsevier Journal of Environmental Psychology, Link: https://www.sciencedirect.com/science/article/abs/pii/S0272494410001015?via%3Dihub, accessed 26.04.2023 & https://drawdown.org/news/insights/the-powerful-role-of-household-actions-in-solving-climate-change, accessed 26.04.2023 & "2018 Climate Change Need Behavior Change", Rare, Link: https://rare.org/report/climate-change-needs-behavior-change/, accessed 26.04.2023 & "The Power of People - Climate Action and the Role of Citizens", The JUMP, 2022, Link: https://takethejump.org/power-of-people, accessed 26.04.2023

119   "The Carbon Majors Database - CDP Carbon Majors Report 2017", Link: https://www.cdp.net/en/articles/media/new-report-shows-just-100-companies-are-source-of-over-70-of-emissions & "The rise in global atmospheric CO_2, surface temperature, and sea level from emissions traced to major carbon producers", Ekwurzel, et al, 2017, Link: https://link.springer.com/article/10.1007/s10584-017-1978-0 & If one analyses the even longer period since the approximate beginning of industrialization (1751-2010), then around 63% of all industrial greenhouse gas emissions are attributable to the 90 largest carbon majors. Source: "Tracing anthropo-

genic carbon dioxide and methane emissions to fossil fuel and cement producers, 1854-2010", R. Heede, Climatic Change, 2013, Link: https://link.springer.com/article/10.1007/s10584-013-0986-y, Zugriff 01.05.2023

120   https://climateaccountability.org/carbon-majors/, accessed 01.05.2023

121   IPCC Assessment Report 6, 2022, Technical Summary (TS-15, Box TS.1): M. Pathak, R. Slade, P.R. Shukla, J. Skea, R. Pichs-Madruga, D. Ürge-Vorsatz,2022: Technical Summary. In: Climate Change 2022: Mitigation of Climate Change. Contribution of Working Group III to the Sixth Assessment Report of the Intergovernmental Panel on Climate Change [P.R. Shukla, J. Skea, R. Slade, A. Al Khourdajie, R. van Diemen, D. McCollum, M. Pathak, S. Some, P. Vyas, R. Fradera, M. Belkacemi, A. Hasija, G. Lisboa, S. Luz, J. Malley, (eds.)]. Cambridge University Press, Cambridge, UK and New York, NY, USA. doi: 10.1017/9781009157926.002 & "COVID curbed carbon emissions in 2020 - but not by much", Tollefson, 2021, Link: https://www.nature.com/articles/d41586-021-00090-3, Zugriff 26.04.2023

122   https://www.imf.org/en/Blogs/Articles/2022/06/30/greenhouse-emissions-rise-to-record-erasing-drop-during-pandemic & https://www.statista.com/statistics/276629/global-co2-emissions/?utm_source=chatgpt.com, accessed 15.01.2025

123   "Young people's climate anxiety revealed in landmark survey", Thompson, 2021, Nature, Link: https://www.nature.com/articles/d41586-021-02582-8 , accessed 25.04.2023

124   "Mapping the Solastalgia Literature: A Scoping Review Study", Galway LP, Beery T, Jones-Casey K, Tasala K., Int J Environ Res Public Health, 2019, doi: 10.3390/ijerph16152662. PMID: 31349659; PMCID: PMC6696016. Link: https://pubmed.ncbi.nlm.nih.gov/31349659/ & https://www.umweltbundesamt.de/themen/gesundheit/umwelteinfluesse-auf-den-menschen/klimawandel-gesundheit/klimawandel-psychische-gesundheit, accessed 25.04.2023

125   Krüger T, Kraus T, Kaifie A. A Changing Home: A Cross-Sectional Study on Environmental Degradation, Resettlement and Psychological Distress in a Western German Coal-Mining Region. Int J Environ Res Public Health. 2022 Jun 10;19(12):7143. doi: 10.3390/ijerph19127143. PMID: 35742391; PMCID: PMC9223024.

126   https://open.spotify.com/episode/5BIiWXkvGJ2bx3Nni5P-KoA?si=3f66580c6f3b4378, min. 27:57, accessed 25.04.2023

127 Many climate scientists could identify with the scene of Jennifer Lawrence's talk show tantrum in the aforementioned film Don't Look Up! The laughter at the Netflix satire stuck in their throats because they found the movie frighteningly real.

128   https://www.un.org/press/en/2022/sgsm21228.doc.htm, accessed 25.04.2023

129   https://www.n-tv.de/politik/1-4-Millionen-demonstrie-ren-in-Deutschland-article21285481.html, accessed 25.04.2023

130   "Positive emotions and climate change", Schneider, et al, 2021, Current Opinion in Behavioral Sciences, Link: https://www.sciencedirect.com/science/article/pii/S2352154621000942, accessed 25.04.2023

131   "A cautionary note about messages of hope: Focusing on progress in reducing carbon emissions weakens mitigation motivation", Hornsey, et al, 2016, Global Environmental Change, Link: https://www.sciencedirect.com/science/article/abs/pii/S0959378016300450 & "Integrating who "we" are with what "we" (will not) stand for: A further extension of the Social Identity Model of Collective Action", van Zomeren, et al, 2018, Link: https://www.tandfonline.com/doi/pdf/10.1080/10463283.2018.1479347?src=getftr, accessed 25.04.2023 & "Affect and emotions as drivers of climate change perception and action: a review", Borsch, 2021, Current Opinion in Behavioral Sciences, Link: https://www.sciencedirect.com/science/article/pii/S2352154621000206, accessed 25.04.2023 & "The Role of Emotion in Global Warming Policy Support and Opposition", Smith, et al, 2013, Link: https://www.researchgate.net/publication/258500851_The_Role_of_Emotion_in_Global_Warming_Policy_Support_and_Opposition, accessed 25.04.2023 & "From anger to action: Differential impacts of eco-anxiety, eco-depression, and eco-anger on climate action and well being", Stanley, 2021, The Journal of Climate Change and Health, Link: https://www.sciencedirect.com/science/article/pii/S2667278221000018, accessed 25.04.2023 & "Why do youth participate in climate activism? A mixed-methods investigation of the #FridaysForFuture climate protests", Haugestad, et al, 2021, Journal of Environmental Psychology Link: https://www.sciencedirect.com/science/aticle/abs/pii/S0272494421001006, accessed 25.04.2023

132   "Dietary behavior as a form of collective action: A social identity model of vegan activism", Judge, et al, Appetite, 2022, Link: https://www.sciencedirect.com/science/article/pii/S0195666321006371, accessed 25.04.2023

133   https://kurier.at/chronik/welt/selbstverbrennung-eines-umweltschuetzers-sorgt-fuer-bewunderung-im-netz/401986307 & https://www.derstandard.at/story/2000077973640/bekannter-us-anwalt-zuendete-sich-wegen-umweltverschmutzung-an & https://www.theguardian.com/environment/2022/may/19/climate-suicides-despair-global-heating, accessed 25.04.2023

134   https://www.tagesschau.de/inland/hungerstreik-klimawandel-101.html & https://www.vienna.at/klimaaktivistin-krumpeck-beendet-hungerstreik-in-wien/7042980 & https://de.wikipedia.org/wiki/Hungern_bis_ihr_ehrlich_seid#:~:text=Bereits%20im%20Vorfeld%20der%20Bundestagswahl,bis%20ihr%20ehrlich%20seid%20angek%C3%BCndigt., accessed 25.04.2023

135   An important note: Anyone who has suicidal thoughts should definitely contact the Samaritans, the German Society for Suicide Prevention or other help services.

136   See for example: https://www.klimareporter.de/gesellschaft/angst-ueberlastung-depression, accessed 25.04.2023

137   "Four Europes: Climate change beliefs and attitudes predict behavior and policy preferences using a latent class analysis on 23 countries", Vachta, et al, 2022, Journal of Environmental Psychology, Link: https://www.sciencedirect.com/science/article/pii/S0272494422000603, accessed 25.04.2023

138   https://www.theguardian.com/environment/2021/feb/27/climatologist-michael-e-mann-doomism-climate-crisis-interview & "Propaganda battle for the climate: How we defeat the instigators of political inaction", Michael E. Mann, 2021 & https://www.theguardian.com/environment/2020/sep/21/meet-the-doomers-some-young-us-voters-have-given-up-hope-on-climate, accessed 25.04.2023

139   https://www.piper.de/buecher/die-klimaschmutzlobby-isbn-978-3-492-31502-9, Zugriff 25.04.2023

140   Quote from an interview for Grist magazine from 26.08.2020, translated into German with DeepL: https://grist.org/energy/footprint-fantasy/, accessed 26.04.2023

141   For example, respondents were asked to assess whether they and their fellow human beings could "achieve something for the protection of nature". As individuals, only just over 40% confirmed this "rather" to "completely". However, when it came to the question of whether we humans or humanity could achieve something together, 67% agreed in a regional context and a full 82% in a global context. Source: https://biologischevielfalt.bfn.de/fileadmin/NBS/images/Veroeffentlichungen/Naturbewusstseinsstudie_2017.pdf, accessed 26.04.2023

142   https://klimakommunikation.klimafakten.de/vor-denken/kapitel-3-frage-dich-wo-sind-deine-ansatzpunkte/, accessed 6/22/22

143   https://open.spotify.com/show/2iprqp419KzbPAExIsofGN, accessed 04/26/2023

144   "We need climate change mitigation and climate change mitigation needs the 'We': a state-of-the-art review of social identity effects motivating climate change action", Masson, et al, 2021, Current Opinion in Behavioral Sciences, Link: https://www.sciencedirect.com/science/article/pii/S2352154621000917, accessed 26.04.2023

145   "Climate change policy support, intended behavior change, and their drivers largely unaffected by consensus messages in Germany", Tschötschel, et al, 2021, Journal of Environmental Psychology, Link: https://www.sciencedirect.com/science/article/pii/S0272494421001080, accessed 26.04.2023

146   "Why We Should Empty Pandora's Box to Create a Sustainable Future: Hope, Sustainability and Its Implications for Education", J. Grund and A. Brock, Sustainability, 2019, Link: https://www.mdpi.com/2071-1050/11/3/893/htm, Zugriff 25.04.2023

147   https://www.bfn.de/sites/default/files/2022-03/Skript620.pdf, accessed 04/26/2023

148   United Nations Framework Convention on Climate Change (UNFCCC) of 1992, Link: https://unfccc.int/resource/docs/convkp/conveng.pdf, accessed 25.04.2023

149   "The role of high-socioeconomic-status people in locking in or rapidly reducing energy-driven greenhouse gas emissions", K. S. Nielsen, et al, Nature Energy, 2021, Link: https://www.nature.com/articles/s41560-021-00900-y, Zugriff 25.04.2023

150   "Confronting Carbon Inequality" by Oxfam, 2020, Link: https://oxfamilibrary.openrepository.com/bitstream/handle/10546/621052/mb-confronting-carbon-inequality-210920-en.pdf & https://wir2022.wid.world/www-site/uploads/2021/12/Summary_WorldInequalityReport2022_German.pdf & "Global carbon inequality over 1990-2019", L. Chancel, Nature Sustainability, 2022, Link: https://www.nature.com/articles/s41893-022-00955-z, Zugriff 26.04.2023

151   https://wir2022.wid.world/www-site/uploads/2022/03/0098-21_WIL_RIM_RAPPORT_A4.pdf & "Impacts of poverty alleviation on national and global carbon emissions", Bruckner, B., Hubacek, K., Shan, Y. et al., Nat Sustain 5, 2022, Link: https://www.nature.com/articles/s41893-021-00842-z, accessed 26.04.2023 & Incidentally, global CO2 emissions would only increase by less than one percentage point if the approximately 650 million people worldwide who have to live on just USD 1.90 a day were freed from their extreme poverty. See the following sources: https://blogs.worldbank.org/opendata/pandemic-prices-and-poverty & https://world-poverty.io/headline, accessed 26.04.2023

152   "Global carbon inequality over 1990-2019", L. Chancel, Nature Sustainability, 2022, Link: https://www.nature.com/articles/s41893-022-00955-z & "World Inequality Report 2022", World Inequality Lab, 2022, Link: https://wir2022.wid.world/www-site/uploads/2022/03/0098-21_WIL_RIM_RAPPORT_A4.pdf & "Confronting Carbon Inequality in the European Union", Oxfam, 2020, Link: https://www.oxfam.org/en/research/confronting-carbon-inequality-european-union, accessed 26.04.2023

153   ibid.

154   https://wir2022.wid.world/www-site/uploads/2021/12/Summary_WorldInequalityReport2022_German.pdf, accessed 04/26/2023

155   "The role of high-socioeconomic-status people in locking in or rapidly reducing energy-driven greenhouse gas emissions", Nielsen,

et al., 2021, Nature, Link: https://www.nature.com/articles/s41560-021-00900-y, accessed 26.04.2023 & "Assessing U.S. consumers' carbon footprints reveals outsized impact of the top 1%", Starr, et al., 2023, Ecological Economics, Link: https://www.sciencedirect.com/science/article/abs/pii/S0921800922003597 & https://www.iea.org/commentaries/the-world-s-top-1-of-emitters-produce-over-1000-times-more-co2-than-the-bottom-1?utm_source=cbnewsletter&utm_medium=email&utm_term=2023-03-04&utm_campaign=This+week+China+s+giant+food+system+Heat+pump+savings+Creating+carbon+space+ & "Carbon Inequality in 2030 - Per capita consumption emissions and the 1.50°C goal", Oxfam, 2021, Link: https://www.oxfamamerica.org/explore/research-publications/carbon-inequality-in-2030/, accessed 26.04.2023

156   "How are high-carbon lifestyles justified? Exploring the discursive strategies of excess energy consumers in the United Kingdom", N. Cass, et al, Energy Research & Social Science, 2023, Link: https://www.sciencedirect.com/science/article/pii/S2214629623000117 & https://www.deutschlandfunk.de/klimabilanz-reiche-rechtfertigungen-100.html, accessed 28.04.2023

157   "Climate Change 2022: Impacts, Adaptation and Vulnerability", Contribution of Working Group II to the Sixth Assessment Report of the IPCC, H.-O. Pörtner, et al., 2022, Link: https://www.ipcc.ch/report/ar6/wg2/ & "Climate Inequality Report 2023 - Fair taxes for a sustainable future in the Global South", World Inequality Lab, L. Chancel, et al., 2023, Link: https://wid.world/wp-content/uploads/2023/01/CBV2023-ClimateInequalityReport-2.pdf, accessed 26.04.2023

158   Dechezleprêtre, A., et al. (2022), "Fighting climate change: International attitudes towards climate policies", OECD Economics Department Working Papers, No. 1714, OECD Publishing, Paris, Link: https://www.oecd-ilibrary.org/economics/fighting-climate-change-international-attitudes-toward-climate-policies_3406f29a-en, accessed 26.04.2023

159   "Scientists' warning on affluence", Wiedmann, et al, 2020, Link: https://www.nature.com/articles/s41467-020-16941-y, accessed 26.04.2023 & "Demographics of emissions", Nature Climate Change, Link: https://www.nature.com/articles/s41558-022-01325-5, Zugriff 26.04.2023

160   See https://www.watson.de/nachhaltigkeit/watson-story/477205579-angst-vor-klimawandel-climaware-gruender-will-mehr-unperfekten-klimaschutz, accessed 26.04.2023

161   See for example "Climate Anxiety", P. Pihkala, MIELI Mental Health Finland, 2019, Link: https://helda.helsinki.fi/bitstream/handle/10138/307626/2019_Pihkala_Climate_Anxiety_report.pdf, accessed 26.04.2023

162   "Climate anxiety in children and young people and their beliefs

about government responses to climate change: a global survey", C. Hickman, et al, The Lancet Planetary Health, 2021, Link: https://www.thelancet.com/journals/lanplh/article/PIIS2542-5196(21)00278-3/fulltext#seccesti-tle10 & "Eco-anxiety in children: A scoping review of the mental health impacts of the awareness of climate change, T. L.-Goodees", et al., 2022, Link: https://www.ncbi.nlm.nih.gov/pmc/articles/PMC9359205/, accessed 26.04.2023

163   ibid.

164   ibid. & see https://www.theguardian.com/lifeandstyle/2019/mar/12/birthstrikers-meet-the-women-who-refuse-to-have-children-until-climate-change-ends, accessed 26.04.2023

165   Google Trends analysis with the search terms "climate anxiety" (worldwide) and "Klimaangst" (Germany) & Twitter searches with the terms #climateanxiety and #klimaangst, accessed 24.04.2023

166   See for example: https://www.droemer-knaur.de/autor/lea-dohm-3009767 & https://www.oekom.de/buch/klima-im-kopf-9783962383817 & https://utopia.de/ratgeber/klimaangst-richtiger-umgang-psychologe/, accessed 26.04.2023

167   https://www.psychologyandglobalhealth.org & American Psychological Association, APA Task Force on Climate Change. (2022) Addressing the Climate Crisis: An Action Plan for Psychologists, Report of the APA Task Force on Climate Change. Link: https://www.apa.org/science/about/publications/climate-crisis-action-plan.pdf, accessed 23.6.22 & 2023, the Rhineland-Palatinate State Chamber of Psychotherapists (LPK RLP) was the first state chamber to publish a corresponding information and demand paper: "KLIMANOTFALL - Auswirkungen ökologischer Krisen auf die psychische Gesundheit", Link: https://www.lpk-rlp.de/fileadmin/user_upload/RZ_LPK_Klimabroschuere_webversion.pdf, accessed 26.04.2023

168   Own translation of the quote "Climate change is a psychological crisis, whatever else it is" by Bruce Poulsen, Psychology Today, 2018, see https://www.psychologistsforfuture.org, accessed 26.04.2023

169   Climaware podcast by Gabriel Baunach and One Pod Wonder, episode "#17 Lea Dohm: How do I overcome the "7 dragons of inaction" and my climate anxiety?", March 26, 2021, Link: https://open.spotify.com/episode/4vA55E6z36ae8fXtZmHnPa, accessed 26.04.2023

170   "Facets of climate anxiety. Psychological foundations of the development of an action-guiding climate awareness", Felix Peter, et al., 2021, Link: https://www.researchgate.net/publication/351814047_Facetten_der_Klimaangst_Psychologische_Grundlagen_der_Entwicklung_eines_handlungsleitenden_Klimabewusstseins_Facets_of_climate_anxiety_Psychological_basics_of_the_development_of_an_action-guiding_clim, accessed 26.04.2023

171   "A cautionary note about messages of hope: Focusing on progress in reducing carbon emissions weakens mitigation motivation", Hornsey, et al., 2016, Link: https://www.sciencedirect.com/science/article/abs/pii/S0959378016300450, accessed 26.04.2023

172   https://www.theguardian.com/environment/2019/jan/25/our-house-is-on-fire-greta-thunberg16-urges-leaders-to-act-on-climate, own translation, accessed 26.04.2023

173   https://www.watson.de/nachhaltigkeit/klimakrise/666050830-angst-vor-der-klimakrise-expertin-verraet-wie-wir-mit-klimaangst-umge-hen-koennen, accessed 04/26/2023

174   "A Scoping Review of Interventions for the Treatment of Eco-Anxiety", P. Baudon and L. Jachenz, 2021, Link: https://www.mdpi.com/1660-4601/18/18/9636, Zugriff 26.04.2023

175   For example, free advice and support services can be found at https://www.psychologistsforfuture.org/unterstuetzung-fuer-engagierte/.

176   In 2018, an environmental psychology study showed that a commitment to sustainability in the community (or an increase in the handprint) makes personal energy saving (i.e. a reduction in the footprint) more likely. This means that even "big" (handprint) actions can trigger the aforementioned domino chain of climate-friendly follow-up actions (positive spillover). Source: "Can community energy initiatives motivate sustainable energy behaviors? The role of initiative involvement and personal pro-environmental motivation", Sloot, et al, Journal of Environmental Psychology, 2018, Link: https://www.sciencedirect.com/science/article/abs/pii/S027249441830330X, accessed 26.04.2023

177   "I did my bit! The impact of electric vehicle adoption on compensatory beliefs and norms in Norway", Nayum, et al., 2022, Energy Research & Social Science, Link: https://www.sciencedirect.com/science/article/pii/S2214629622000482 & "Experiences of pride, not guilt, predict pro-environmental behavior when pro-environmental descriptive norms are more positive", Bissing-Olson, et al., 2016, Journal of Environmental Psychology, Link: https://www.sciencedirect.com/science/article/abs/pii/S0272494416300019?via%3Dihub, accessed 26.04.2023

178   "Sternstunde der Philosophie" by SRF-Kultur from October 17, 2021, Link: https://www.youtube.com/watch?v=9lh0YuuGr00, accessed 26.04.2023

179   "Demand-side solutions to climate change mitigation consistent with high levels of well-being", F. Creutzig, et al, Nature Climate Change, 2021, Link: https://www.nature.com/articles/s41558-021-01219-y, Zugriff 26.04.2023

180   https://twitter.com/theCCCuk/sta-tus/1123838907999752192?s=20&t=HO5j4rPqMAfnn3GqTMjjdA, Zugriff 26.04.2023

181   See, for example, Climate Change Needs Behavior Change: Making the Case For Behavioral Solutions to Reduce Global Warming. Williamson, K., Satre-Meloy, A., Velasco, K., & Green, K., 2018, Arlington, VA: Rare. Link: https://rare.org/wp-content/uploads/2019/02/2018-CCN-BC-Report.pdf & Emissions Gap Report 2020. United Nations Environment Programme (2020). Nairobi. Link: https://www.unep.org/emissions-gap-report-2020 & Climate Change 2022: Mitigation of Climate Change. Contribution of Working Group III to the Sixth Assessment Report of the Intergovernmental Panel on Climate Change, P.R. Shukla, et al., IPCC, 2022, Cambridge University Press, Cambridge, UK and New York, NY, USA. doi: 10.1017/9781009157926, accessed 26.04.2023

182   "Decay radius of climate decision for solar panels in the city of Fresno, USA", K. Barton-Henry, et al., 2021, Link: https://www.nature.com/articles/s41598-021-87714-w, Zugriff 26.04.2023

183   "Analysis: Geographical and temporal differences in the approval of climate protection policies in Germany", S. Levi, et al, Ariadne Project, 2023, Link:
 https://ariadneprojekt.de/publikation/analyse-geographische-und-zeitliche-unterschiede-in-der-zustimmung-zu-klimaschutzpolitik-in-deutschland/, accessed 04/26/2023

184   https://theconversation.com/climate-change-yes-your-individual-action-does-make-a-difference-115169 & https://theconversation.com/how-world-leaders-high-carbon-travel-choices-could-delay-climate-action-162784, accessed 26.04.2023

185   See for example "The Climate Progress Survey" by the World Economic Forum (WEF), 2021, Link: https://www3.weforum.org/docs/SAP_WEF_Sustainability_Report.pdf, accessed 15.10.22

186   "Statements about climate researchers' carbon footprints affect their credibility and the impact of their advice", Attari, et al, 2016, Link: https://link.springer.com/article/10.1007/s10584-016-1713-2, Zugriff 26.04.2023

187   "Climate change communicators' carbon footprints affect their audience's policy support", Attari, et al, 2019, Link: https://link.springer.com/article/10.1007/s10584-019-02463-0, Zugriff 26.04.2023

188   Center for Research on Environmental Decisions, Columbia University in New York City, "The Psychology of Climate Change Communication", Link: http://guide.cred.columbia.edu/guide/sec4.html, accessed 6.4.2022 & "Did concern about COVID-19 drain from a 'finite pool of worry' for climate change? Results from longitudinal panel data", Gregersen, et al, 2022, The Journal of Climate Change and Health, Link: https://www.sciencedirect.com/science/article/pii/S2667278222000335 & "Effect of "finite pool of worry" and COVID-19 on UK climate change percep-

tions", Evensen, et al, 2021, PNAS, Link: https://www.pnas.org/doi/10.1073/pnas.2018936118, Zugriff 26.04.2023

189   "A Finite Pool of Worry or a Finite Pool of Attention? Evidence and Qualifications", Sisco, et al. 2020, Link: https://www.researchgate.net/publication/346586727_A_Finite_Pool_of_Worry_or_a_Finite_Pool_of_Attention_Evidence_and_Qualifications & on the connection with the climate crisis, see also "Effect of "finite pool of worry" and COVID-19 on UK climate change perceptions", Evensen, et al., PNAS, 2021, Link: https://www.pnas.org/doi/10.1073/pnas.2018936118 & "Climate Change + Consumer Behavior. A Global Advisor Survey", Ipsos, 2021, Link: https://www.ipsos.com/en/climate-change-consumer-behaviour-2021, accessed 26.04.2023

190   https://www.umweltbundesamt.de/themen/big-points-klimafreundliche-konsumentscheidungen, accessed 04/26/2023

191   See IPCC, 2022: Climate Change 2022: Mitigation of Climate Change. Contribution of Working Group III to the Sixth Assessment Report of the Intergovernmental Panel on Climate Change, P.R. Shukla, J. Skea, R. Slade, et al, Cambridge University Press, Cambridge, UK and New York, NY, USA, Chapter 5, Figure 5.8, p. 532. doi: 10.1017/9781009157926

192   There are actually two other questions that have a major impact on personal greenhouse gas emissions. Firstly, how many dogs, cats, horses etc. you get: giving up pets saves around 0.8 tons of greenhouse gases per year. And secondly, how many children you have: having one child less saves an average of 58.6 tons of greenhouse gases per year in industrialized nations. But because this is ethical black ice, I won't go into any more detail.

193   ibid. & "The climate mitigation gap: education and government recommendations miss the most effective individual actions", S. Wynes and K. A. Nicholas, Environ. Res. Lett. 12, 2017, link: https://iopscience.iop.org/article/10.1088/1748-9326/aa7541/meta & "Valuing the Multiple Impacts of Household Food Waste", M. von Massow, et al., 2019, Link: https://www.frontiersin.org/articles/10.3389/fnut.2019.00143/full & DGNB study on CO2 emissions from buildings, Link: https://www.dgnb.de/de/aktuell/pressemitteilungen/2021/studie-co2-emissionen-bauwerke, accessed 26.04.2023

194   https://klimakommunikation.klimafakten.de/showtime/kapitel-9-bleibe-positiv-sowohl-im-ton-wie-im-inhalt/ & https://www.chelseagreen.com/product/what-we-think-about-when-we-try-not-to-think-about-global-warming/, accessed 26.04.2023

195 See for example https://www.axios.com/unreliable-news-sources-social-media-engagement-297bf046-c1b0-4e69-9875-05443b1dca73.html, accessed 26.04.2023

196   "The spread of true and false news online", Vosoughi, et al,

2018, Science, Link: https://www.science.org/doi/10.1126/science.aap9559, accessed 26.04.2023

197   https://www.theguardian.com/environment/2019/oct/10/fossil-fuel-firms-social-media-fightback-against-climate-action, accessed 04/26/2023 https://www.theguardian.com/technology/2020/feb/21/climate-tweets-twitter-bots-analysis & Draft Conference Paper: https://cssn.org/wp-content/uploads/2020/12/DRAFT-Twitter_Discourses_on_Climate_Change-20200224.pdf & "Bots and online climate discourses: Twitter discourse on President Trump's announcement of U.S. withdrawal from the Paris Agreement", Marlow, et al., 2022, Link: https://www.tandfonline.com/doi/abs/10.1080/14693062.2020.1870098, Zugriff 26.04.2023

198   https://www.theguardian.com/environment/2019/oct/10/fossil-fuel-firms-social-media-fightback-against-climate-action, accessed 04/26/2023

199   https://climatecommunication.yale.edu/publications/climate-spiral-silence-america/, accessed 26.04.2023

200   "Climate of silence: Pluralistic ignorance as a barrier to climate change discussion", Geiger, et al, 2016, Journal of Environmental Psychology, Link: https://www.sciencedirect.com/science/article/abs/pii/S027249441630038X & https://www.klimafakten.de/meldung/warum-viele-menschen-lieber-nicht-ueber-den-klimawandel-sprechen, accessed 26.04.2023

201   https://noelle-neumann.de/wissenschaftliches-werk/schweigespirale, accessed 04/26/2023

202 Planetary Health Action Survey, University of Erfurt, 2022, Link: https://projekte.uni-erfurt.de/pace/_files/PACE_W07-09.pdf#page=24 & briq policy monitor, 2022, Link: https://news.briq-institute.org/de/2022/09/01/bereitschaft-der-deutschen-zum-klimaschutz/ & see https://www.klimareporter.de/gesellschaft/verzerrte-wahrnehmung, accessed 26.04.2023

203   For the entire section, see podcast "Über Klima sprechen" by klimafakten.de and Gabriel Baunach, introductory episode, link: https://open.spotify.com/episode/0gps27vJh2Pjtuc4LbdxoF?si=b8f598c64f50432c, accessed 26.04.2023

204   https://www.klimafakten.de/meldung/das-einfachste-mittel-gegen-den-klimawandel-im-freundes-und-familienkreis-darueber-reden, accessed 04/26/2023

205   Quote from the podcast "How to save a planet", Ep. "Is your Carbon Footprint BS?" from 28.03.2021, Link: https://open.spotify.com/episode/5LlRVInqBXtZgEzWTHOaTg?si=8e4f7c4cbcee47e5, accessed 01.05.2023

206   Climaware Podcast, Gabriel Baunach and One Pod Wonder, Ep.

#9 with Stefan Rahmstorf, Link: https://open.spotify.com/episode/4NeZx-InOnpjrTI8sooSEhq?si=0a463908363940e7, from min. 40:50, accessed 01.05.2023

207   Webster, R. & Marshall, G. (2019) The #TalkingClimate Handbook. How to have conversations about climate change in your daily life. Oxford: Climate Outreach, Link: https://climateoutreach.org/reports/how-to-have-a-climate-change-conversation-talking-climate/, accessed 26.04.2023

208   See, for example, Epley, Nicholas, and Thomas Gilovich. 2016. "The Mechanics of Motivated Reasoning." Journal of Economic Perspectives, 30 (3): 133-40. link: https://www.aeaweb.org/articles?id=10.1257/jep.30.3.133, Zugriff 27.04.2023

209   https://www.klimafakten.de/meldung/p-l-u-r-v-dies-sind-die-haeufigsten-desinformations-tricks-von-wissenschafts-leugnern, accessed 04/26/2023

210   "Discourses of climate delay", Lamb, et al, 2020, Link: https://www.cambridge.org/core/journals/global-sustainability/article/discourses-of-climate-delay/7B11B722E3E3454BB6212378E32985A7 & see https://www.klimafakten.de/meldung/nicht-ich-nicht-jetzt-nicht-so-zu-spaet-mit-welchen-argumentationsmustern-klimaschutz, accessed 26.04.2023

211 "Children can foster climate change concern among their parents", Lawson, et al, 2019, Link: https://www.nature.com/articles/s41558-019-0463-3 & see https://www.klimafakten.de/meldung/kinder-koennen-das-klimabewusstsein-ihrer-eltern-deutlich-beeinflussen-vor-al-lem-toechter, accessed 26.04.2023

212   https://klimakommunikation.klimafakten.de/nachwort-klimak-ommunikation-allein-genuegt-nicht/, accessed 04/26/2023

213   https://en.wikipedia.org/wiki/Economy_of_Poland & "Banking on Climate Chaos 2024 - Fossil Fuel Finance Report 2024", Link: https://www.bankingonclimatechaos.org/wp-content/uploads/2024/09/BOCC_2024_SUMMARY.pdf, accessed 22.01.2025

214   https://press.un.org/en/2022/sgsm21228.doc.htm, accessed 04/26/2023

215   "Banking on Climate Chaos 2022 - Fossil Fuel Finance Report 2022", Link: https://www.bankingonclimatechaos.org/#data-panel, accessed 26.04.2023 & sample analysis by investigative journalists on "Sustainable Finance" at HSBC: https://www.thebureauinvestigates.com/stories/2022-10-31/mines-pipelines-and-oil-rigs-what-hsbcs-sustainable-finance-really-pays-for, accessed 26.04.2023

216   https://www.bbc.com/news/science-environment-63975173, accessed 04/26/2023

217   https://www.theguardian.com/money/article/2024/jun/15/ethi-

cal-banking-in-the-uk-how-to-put-your-everyday-account-to-good-use, accessed 22.01.2025See also "German banks must pick up speed - Sustainability analysis of German banks", WWF Germany, 2021, Link: https://www.wwf.de/fileadmin/fm-wwf/Publikationen-PDF/Unternehmen/WWF-Zweites-Bankenrating.pdf, accessed 26.04.2023

218   https://makemymoneymatter.co.uk/wp-content/uploads/2024/02/Make-My-Money-Matter-Climate-Action-Report-2024.pdf, accessed 22.01.2025

219   https://cps.org.uk/wp-content/uploads/2023/07/CPS-Retail-Therapy-FINAL.pdf, accessed 22.01.2025

220   See also SWR documentary "How banks trick sustainable ETFs" from 6.9.22, link: https://www.swrfernsehen.de/vollbild-recherchen-die-mehr-zeigen/so-tricksen-banken-bei-nachhaltigen-etf-100.html, accessed 01.05.2023

221   "Climate Funds: Are They Paris Aligned?", InfluenceMap, 2021, Link: https://influencemap.org/report/Climate-Funds-Are-They-Paris-Aligned-3eb83347267949847084306dae01c7b0, Zugriff 26.04.2023

222   https://press.un.org/en/2022/sgsm21228.doc.htm, translated with DeepL.com/translator, accessed 26.04.2023

223   "Climate Funds: Are They Paris Aligned?", InfluenceMap, 2021, Link: https://influencemap.org/report/Climate-Funds-Are-They-Paris-Aligned-3eb83347267949847084306dae01c7b0, Zugriff 26.04.2023

224   See for example https://wiwin.de, accessed 26.04.2023

225   "What boards should know about ESG developments in the 2021 proxy season", EY Center for Board Matters, 2021, Link: https://higherlogicdownload.s3.amazonaws.com/GOVERNANCEPROFESSIONALS/a8892c7c-6297-4149-b9fc-378577d0b150/UploadedImages/ey-what-boards-should-know-about-esg-developments-2021-proxy-season-cbm-1.pdf & https://www.cfodive.com/news/shareholder-support-surges-action-climate-change-ey/604605/, accessed 26.04.2023

226   See https://www.fridel.de, accessed 26.04.2023

227   "Crop-damaging temperatures increase suicide rates in India", T. A. Carleton, PNAS, 2017, Link: https://www.pnas.org/doi/10.1073/pnas.1701354114#bibliography, Zugriff 01.05.2023

228   Climaware Podcast Ep. 13, 2021, from Min. 37:30, Link: https://open.spotify.com/episode/1dzNHwwZNiMLb9QMavsvH6?si=8c9d7a0d-9ecd4cf3, accessed 26.04.2023

229   https://www.euronews.com/green/2022/05/24/how-can-the-class-of-2022-prevent-climate-havoc-be-part-of-the-solution-says-un-chief, translated with DeepL.com/translator, accessed 26.04.2023

230   See https://www.theguardian.com/environment/2021/sep/06/gen-z-climate-change-careers-jobs, accessed 26.04.2023

231   https://climatemind.de, accessed 01.05.2023

232   E.g. for the building sector: https://www.reuters.com/sustainability/climate-energy/long-ambition-short-people-how-skills-gap-could-scupper-uks-bid-decarbonise-2024-11-28/?utm_source=chatgpt.com, accessed 22.01.2025

233   "Net Zero by 2050 - A Roadmap for the Global Energy Sector", IEA, 2021, Link: https://www.iea.org/reports/net-zero-by-2050, accessed 26.04.2023

234   https://news.sky.com/story/theyre-ignoring-all-the-alarms-contractor-resigns-from-shell-with-warning-to-staff-about-extreme-harm-to-planet-12619719 & https://www.linkedin.com/posts/caroline-dennett-6161a814_jumpship-truthteller-activity-6934409781495431168-7lIf?utm_source=share&utm_medium=member_desktop, translated with DeepL.com/translator, 26.04.2023

235   https://2020.de/do/ & https://www.linkedin.com/in/daniel-obst/, 26.04.2023

236   https://www.janine-steeger.de & https://www.bne-portal.de/bne/de/bne-jetzt/unterseiten/Testimonials/Janine_Steeger/Janine-Steeger.html, accessed 26.04.2023

237   https://www.rtl.de/cms/sendungen/news/klima-update/, accessed 04/26/2023

238   https://www.oekom.de/buch/going-green-9783962381769, Zugriff 26.04.2023

239   See for example https://www.bloomberg.com/news/features/2023-01-05/how-to-quit-your-job-to-fight-climate-change, accessed 01.05.2023

240   https://www.mckinsey.com/capabilities/people-and-organizational-performance/our-insights/european-talent-is-ready-to-walk-out-the-door-how-should-companies-respond, accessed 04/26/2023

241   See https://konzeptwerk-neue-oekonomie.org/bausteine-fuer-klimagerechtigkeit/arbeitszeitverkuerzung/, accessed 26.04.2023

242   I'm thinking here of a quote from the US radio presenter and financial book author Dave Ramsey: "We buy things we don't need with money we don't have to impress people we don't like."

243   https://autonomy.work/portfolio/icelandsww/ & https://autonomy.work/wp-content/uploads/2023/02/The-results-are-in-The-UKs-four-day-week-pilot.pdf, accessed 26.04.2023

244   What One Person Can Do About Climate Change", Ella Lagé, TEDxHamburg, Link: https://www.youtube.com/watch?v=LRQWXFCaOGs&t=403s, accessed 04/26/2023

245   https://gofossilfree.org/de/press-release/kohle-ol-und-gas-tabu-berlin-zieht-offentliche-gelder-von-klimasundern-ab/ & https://divest-mentdatabase.org, accessed 26.04.2023

246   "Beyond national climate action: the impact of region, city, and business commitments on global greenhouse gas emissions", Kuramochi, et al., 2019, Link: https://www.tandfonline.com/doi/full/10.1080/14693062.2020.1740150, Zugriff 26.04.2023

247   "Winning the Race to Net Zero: The CEO Guide to Climate Advantage", World Economic Forum, Insight Report, 2022, Link: https://www3.weforum.org/docs/WEF_Winning_the_Race_to_Net_Zero_2022.pdf, translated with DeepL.com/translator, accessed 26.04.2023

248   https://www.linkedin.com/feed/update/urn:li:activity:6896732527684636672?updateEntityUrn=urn%3Ali%3Afs_feedUp-date%3A%28V2%2Curn%3Ali%3Aactivity%3A6896732527684636672%29, Zugriff 26.04.2023

249   https://ghgprotocol.org, accessed 04/26/2023

250   ibid.

251   See https://www.cleanenergywire.org/news/corporate-net-zero-commitments-send-ripple-effects-across-globe, accessed 26.04.2023

252   https://sciencebasedtargets.org/how-it-works, accessed 04/26/2023

253   https://www.theguardian.com/environment/2023/jan/18/re-vealed-forest-carbon-offsets-biggest-provider-worthless-verra-aoe & https://www.zeit.de/2023/04/co2-zertifikate-betrug-emissionshandel-kli-maschutz?utm_referrer=https%3A%2F%2Fwww.linkedin.com%2F, accessed 26.04.2023

254   "The fairy tale of climate-neutral products", vzbv position paper, November 2022, Link: https://www.vzbv.de/pressemitteilungen/vzbv-fordert-verbot-von-werbung-mit-klimaneutralitaet & https://www.wettbewerbszentrale.de/de/_pressemitteilungen/?id=385 & https://www.klimafakten.de/meldung/kontrolle-status-ein-gutes-gewissen-wieso-greif-en-verbraucherinnen-zu-klimaneutralen, accessed 26.04.2023

255   https://www.maersk.com/news/articles/2021/02/17/maersk-first-carbon-neutral-liner-vessel-by-2023 & https://www.france24.com/en/live-news/20230914-maersk-unveils-world-s-first-bio-methanol-container-ship-1 & https://www.ssab.com/en/fossil-free-steel, accessed 26.04.2023

256   https://www.ikea.com/at/de/this-is-ikea/sustainable-everyday/zweites-leben-shop-pub076edc20, accessed 04/26/2023

257   https://toogoodtogo.de/de/, accessed 04/26/2023

258   https://www.globalbattery.org, accessed 04/26/2023

259   See for example https://www.pro-bono-camp.org/das-projekt/, accessed 26.04.2023

260   https://www.patagonia.com/ownership/, accessed 04/26/2023

261   https://de.blog.ecosia.org/ecosia-finanzberichte-baumplanzbele-ge/, accessed 01.05.2023

262   https://info.ecosia.org/what, accessed 12/14/22

263   https://influencemap.org/briefing/Responsible-Climate-Poli-cy-Engagement-Leadership-in-Action-29941 & "Corporate Climate Policy Footprint: The 25 most influential companies blocking climate policy action globally", InfluenceMap, 2022, Link: https://influencemap.org/report/Corporate-Climate-Policy-Footprint-2022-20196, accessed 26.04.2023

264   "Climate Solutions at Work: An Employee Guide to Drawdown-Aligned Business", Project Drawdown, 2021, Link: https://www.drawdown.org/sites/default/files/210920_Drawdown_AtWork_06.pdf, accessed 26.04.2023

265   https://www.theguardian.com/technology/2019/sep/10/jeff-bezos-amazon-climate-strike-aecj & https://www.reuters.com/business/sustainable-business/society-watch-how-employees-are-taking-their-companies-task-over-climate-change-2022-04-18/, accessed 26.04.2023

266   https://amazonemployees4climatejustice.medium.com/public-letter-to-jeff-bezos-and-the-amazon-board-of-directors-82a8405f5e38, accessed 04/26/2023

267   https://techworkerscoalition.org/climate-strike/, accessed 04/26/2023

268   An example of such a sustainability education program for employees: https://kiteinsights.com/climate-school-sign-up/, accessed 26.04.2023

269   "Climate Solutions at Work: An Employee Guide to Drawdown-Aligned Business", Project Drawdown, 2021, Link: https://www.drawdown.org/sites/default/files/210920_Drawdown_AtWork_06.pdf, accessed 26.04.2023

270   The NGO Germanwatch, which originally developed the hand-print concept, does not distinguish between social and political handprints, as it considers all social engagement to be political.

271   https://klimavoracht.de/programmdaten/ & https://www.ard-media.de/media-perspektiven/fachzeitschrift/2020/detailseite-2020/der-klimawandel-im-oeffentlich-rechtlichen-fernsehen/ & https://taz.de/Send-ezeit-fuers-Klima/!5914530/ & see https://www.klimafakten.de/meldung/aufmerksamkeit-medienlogiken-kernbotschaften-was-wir-aus-der-pande-mie-lernen-koennen, accessed 26.04.2023

272   https://www.theguardian.com/environment/2021/sep/15/cake-mentioned-10-times-more-than-climate-change-on-uk-tv-report, accessed 23.01.2025

273   https://reutersinstitute.politics.ox.ac.uk/climate-change-and-

news-audiences-report-2024-analysis-news-use-and-attitudes-eight-countries, accessed 07.02.2025

274   https://reutersinstitute.politics.ox.ac.uk/climate-change-and-news-audiences-report-2024-analysis-news-use-and-attitudes-eight-countries, accessed 07.02.2025

275   https://www.oekom.de/buch/medien-in-der-klima-krise-9783962383855 & "Review: Ecological awareness, anxiety, and actions among youth and their parents - a qualitative study of newspaper narratives", L. Benoit, et al., 2022, Link: https://acamh.onlinelibrary.wiley.com/doi/epdf/10.1111/camh.12514?casa_token=ajfUsaTSPjcAAAAA%3A7q8Qc-dKVl9iPbGol4w1DcaUuub-LZvqMZOcDVKg4Hd8--sWhJQb-zTp-fu3N5FsCo9vAH2P_moqaA, accessed 26.04.2023

276   https://medienleitfaden-klima.de, accessed 04/26/2023

277   https://reutersinstitute.politics.ox.ac.uk/oxford-climate-journalism-network, accessed 23.01.2025

278   "The (Un)political Perspective on Climate Change in Education-A Systematic Review", J. Kranz, et al., 2022, Link: https://www.mdpi.com/2071-1050/14/7/4194 & see https://www.klimafakten.de/meldung/politik-der-blinde-fleck-der-klimabildung, accessed 26.04.2023

279   By this I mean "non-formal education". See https://www.ewi-psy.fu-berlin.de/einrichtungen/weitere/institut-futur/Projekte/Dateien/Brock_-A_-Grund_-J_2020_Non-formale_BNE_Divers_volatil_und_dabei_feste1.pdf, accessed 26.04.2023

280   "The (Un)political Perspective on Climate Change in Education-A Systematic Review", Kranz, et al., 2022, Link: https://www.mdpi.com/2071-1050/14/7/4194 & "The climate mitigation gap: education and government recommendations miss the most effective individual actions", Wynes, 2017, Link: https://iopscience.iop.org/article/10.1088/1748-9326/aa7541, Zugriff 26.04.2023

281   For the entire paragraph: https://www.theguardian.com/world/2022/nov/12/barcelona-students-to-take-mandatory-climate-crisis-module-from-2024, accessed 26.04.2023

282   https://www.dataforprogress.org/memos/accountable-allies-the-undue-influence-of-fossil-fuel-money-in-academia & https://fossilfreeresearch.org, accessed 01.05.2023

283   https://www.ox.ac.uk/news/2020-04-27-oxford-announces-historic-commitment-fossil-fuel-divestment, accessed 31.01.2025

284   https://www.cam.ac.uk/notices/news/the-university-and-funding-from-fossil-fuel-companies, accessed 28.01.2025

285   See "Whole Institution Approach". See for example https://netzwerk-n.org/wp-content/uploads/2022/06/Positionspapier-ausfuehrliche-Version-1.pdf & https://www.germanwatch.org/de/20183 & for positive

examples see https://netzwerk-n.org/ressourcen/good-practice/, accessed 26.04.2023

286   Climaware podcast episode #23, 2021: https://open.spotify. com/episode/7KRqdqja57IEPqbsaTheea?si=73da11ebbd40434b, accessed 26.04.2023

287   https://www.dieklimawette.de, accessed 04/26/2023

288   "Criteria for assessing the transformation potential of sustainability initiatives", Federal Environment Agency, 2019, Link: https:// www.bmuv.de/fileadmin/Daten_BMU/Pools/Forschungsdatenbank/ fkz_3714_17_100_nachhaltiges_handeln_bf.pdf, accessed 26.04.2023

289   https://commonslibrary.parliament.uk/research-briefings/cdp-2024-0086/, accessed 28.01.2025

290   https://difu.de/publikationen/2020/aktiv-fuer-den-klima-schutz-wie-sie-als-sportverein-profitieren, accessed 04/26/2023

291   https://www.vatican.va/content/francesco/de/encyclicals/doc-uments/papa-francesco_20150524_enciclica-laudato-si.html, accessed 04/26/2023

292   https://www.churchofengland.org/resources/net-zero-car-bon-routemap, accessed 28.01.2025

293   Of course, this and everything else also applies to Muslim, Jewish, Buddhist and all other religious or spiritual associations and communities, even if I do not explicitly mention them here.

294   https://klima-kollekte.de/fileadmin/user_upload/UBA_Handre-ichung_Wie-man-beginnen-kann---Umwelt-und-Klimaschutz-in-Kirchen-gemeinden.pdf, accessed 04/26/2023

295   https://www.kath.ch/newsd/studie-franziskus-koennte-kli-ma-beeinflussen/, accessed 04/26/2023

296   https://web.de/magazine/wirtschaft/vermoegen-ver-fuegt-kirche-deutschland-33621932 & https://kontrast.at/vermoe-gen-katholische-kirche-oesterreich/ & https://www.spiegel.de/wirtschaft/ soziales/katholische-kirche-das-versteckte-vermoegen-der-bistue-mer-a-1269846.html, accessed 26.04.2023

297   There are huge differences in salaries, for example when comparing systemically relevant professions such as nursing with the top salaries of managers. In addition, low-taxed inheritances worth millions, such as transfers of company shares, lead to an unfair concentration of power and wealth. This means that widespread beliefs such as "If you work a lot, you make a lot of money", "If you have a lot of money, you have worked a lot" and "All citizens contribute their fair share to the common good through taxes" reflect reality less and less.

298   https://www.activephilanthropy.org/de/, accessed 04/26/2023

299   https://www.unep.org/news-and-stories/press-release/world-

needs-usd-81-trillion-investment-nature-2050-tackle-triple, accessed 04/26/2023

300   https://getactive.earth/projects, accessed 04/26/2023

301   https://www.effectivealtruism.org, accessed 04/26/2023

302   https://ssir.org/articles/entry/protest_movements_could_be_more_effective_than_the_best_charities, accessed 04/26/2023

303   https://www.socialchangelab.org/_files/ug-d/503ba4_052959e2ee8d4924934b7efe3916981e.pdf, Zugriff 26.04.2023

304   https://www.catf.us & https://founderspledge.com/stories/the-clean-air-task-force-high-impact-funding-opportunity & https://founderspledge.com/research/fp-climate-change & https://docs.google.com/spreadsheets/d/1q6srpmt5VkdXLGfYzqHqkU3hvGUwPKjA67uxqYI0Upw/edit#gid=0, accessed 26.04.2023

305   https://www.socialchangelab.org & https://www.givinggreen.earth/top-climate-change-nonprofit-donations-recommendations & https://founderspledge.com/funds/climate-change-fund & https://effektiv-spenden.org/effektiver-klimaschutz/ & see also https://utopia.de/ratgeber/effektiver-altruismus-so-kannst-du-besser-helfen/, accessed 26.04.2023

306   https://www.climateemergencyfund.org/ourgrantees, accessed 01.05.2023

307   At 400 Gt CO2-Budget from 2020 and 67% probability. See https://www.mcc-berlin.net/en/research/co2-budget.html, accessed 27.04.2023

308   https://news.un.org/en/story/2022/07/1123482 & https://digitallibrary.un.org/record/3982508?ln=en, accessed 27.04.2023

309   Basic Law Article 20a, Link: https://www.gesetze-im-internet.de/gg/art_20a.html, accessed 27.04.2023

310   "Global trends in climate change litigation: 2022 snapshot", J. Setzer & C. Higham, 2022, Grantham Research Institute on Climate Change and the Environment and the Centre for Climate Change Economics and Policy. Link: https://www.cccep.ac.uk/wp-content/uploads/2022/06/Global-trends-in-climate-change-litigation-2022-snapshot.pdf & for some examples see "Climate change litigation" databases, Link: http://climatecasechart.com, accessed 27.04.2023

311   https://www.vanuatuicj.com/resolution & https://www.sprep.org/news/vanuatu-implores-world-leaders-to-vote-for-international-court-of-justice-climate-resolution & https://www.theguardian.com/environment/2023/mar/29/united-nations-resolution-climate-emergency-vanuatu, accessed 27.04.2023

312   https://www.klimareporter.de/gesellschaft/neue-verfassungsklage-gegen-klimagesetz & https://www.klimareporter.de/deutschland/klage-gegen-1-8-milliarden-tonnen-co2 & https://www.klimareporter.

de/international/gerichte-treffen-weitreichende-klimaentscheidungen, accessed 27.04.2023

313 https://insideclimatenews.org/news/17012020/climate-change-fossil-fuel-company-lawsuits-timeline-exxon-children-california-cities-attorney-general/, accessed 27.04.2023

314 "Assessing ExxonMobil's global warming projections", Supran, et al, 2022, Science, Link: https://www.science.org/doi/10.1126/science.abk0063, accessed 27.04.2023

315 https://climate-laws.org/geographies/germany/litigation_cases/luciano-lliuya-v-rwe & https://www.germanwatch.org/de/der-fall-rwe & 2022 a group from Cologne even filed criminal charges against RWE Power AG for homicide caused by the mining and burning of lignite. Source: https://www.kritischeaktionaere.de/rwe/strafanzeige-wegen-toetungsdelikten-gegen-die-rwe-power-ag/, accessed 20.1.23

316 https://climateemergencydeclaration.org/climate-emergency-declarations-cover-15-million-citizens/, accessed 28.01.2025

317 https://maps.transitionnetwork.org/, accessed 28.01.2025

318 Climaware Podcast, Gabriel Baunach and One Pod Wonder, Ep. #21, 2021, Link: https://open.spotify.com/episode/5fxMg2plxFnaM7UooHBf7Q, accessed 27.04.2023

319 https://www.theguardian.com/environment/2024/nov/15/coal-oil-and-gas-lobbyists-granted-access-to-cop29-says-report, accessed 28.01.2025

320 "Can citizen pressure influence politicians' communication about climate change? Results from a field experiment", S. Wynes, et al., 2021, Link: https://link.springer.com/content/pdf/10.1007/s10584-021-03215-9.pdf, Zugriff 27.04.2023

321 However, approval of various climate protection measures varies greatly from region to region. For example, the Ariadne project 2023 was able to demonstrate a very strong urban-rural divide. Source: "Analysis: Geographical and temporal differences in approval of climate protection policy in Germany", S. Levi, et al., 2023, Link: https://ariadneprojekt.de/publikation/analyse-geographische-und-zeitliche-unterschiede-in-der-zustimmung-zu-klimaschutzpolitik-in-deutschland/, accessed 01.05.2023

322 M. Pathak, R. Slade, P.R. Shukla, J. Skea, R. Pichs-Madruga, D. Ürge-Vorsatz,2022: Technical Summary. In: Climate Change 2022: Mitigation of Climate Change. Contribution of Working Group III to the Sixth Assessment Report of the Intergovernmental Panel on Climate Change [P.R. Shukla, J. Skea, R. Slade, A. Al Khourdajie, R. van Diemen, D. McCollum, M. Pathak, S. Some, P. Vyas, R. Fradera, M. Belkacemi, A. Hasija, G. Lisboa, S. Luz, J. Malley, (eds.)]. Cambridge University Press, Cambridge,

UK and New York, NY, USA. doi: 10.1017/9781009157926.002, Link: https://www.ipcc.ch/report/sixth-assessment-report-working-group-3/, accessed 01.05.2023

323   "Why Civil Resistance Works", Erica Chenoweth and Maria J. Stephan, 2012, Link: https://cup.columbia.edu/book/why-civil-resistance-works/9780231156820 & "Why Civil Resistance Works. The Strategic Logic of Nonviolent Conflict", Erica Chenoweth and Maria J. Stephan, International Security, Vol. 33, Issue 1, 2008, accessed 27.04.2023

324   https://www.faz.net/aktuell/wirtschaft/nach-klima-strategie-einigung-merkel-verteidigt-klimapaket-16393990.html, accessed 04/27/2023

325   Climaware Podcast, Gabriel Baunach and One Pod Wonder, Ep. #22, 2021, Link: https://open.spotify.com/episode/5NqhgiieK3WODkGhCHcHLF?si=42fd98c3490c4d7e, accessed 27.04.2023

326   See "Theory of change - Creating a social mandate for climate action", Climate Outreach, 2020, Link: https://climateoutreach.org/reports/theory-of-change/ & https://www.climate-handprint.de/wissen/unsere-motivation/, accessed 27.04.2023

327   https://www.bbc.com/news/world-europe-50694361, accessed 27.04.2023

328   "Civil Disobedience", Henry David Thoreau, [What does it all mean?], Reclams Universal-Bibliothek Volume 19053, Reclam, Philipp, 2013

329   There are various definitions of civil disobedience, including by well-known sociological and philosophical scholars such as Hannah Arendt, John Rawls and Jürgen Habermas. Sources: https://www.bpb.de/shop/zeitschriften/apuz/138281/ziviler-ungehorsam-annaeherung-an-einen-umkaempften-begriff/ & anthology "Ziviler Ungehorsam im Rechtsstaat", J. Habermas, et al., ed. Peter Glotz, 1983, Suhrkamp & "Theorie der Gerechtigkeit", J. Rawls, 1975 & "Widerstand oder Terrorismus? A theoretical foundation", F. Höntzsch, 2021, Link: https://link.springer.com/article/10.1007/s11615-020-00290-y#Bib1 & see https://www.klimareporter.de/protest/wie-geht-es-weiter-mit-dem-protest-simon-teune, accessed 27.04.2023

330   See https://www.liberties.eu/de/stories/was-ist-ziviler-ungehorsam-definition-beispiele/44569, accessed 27.04.2023

331   "Breaking laws for the climate? | angel asks | Documentaries & reports", hrfernsehen, Link: https://www.youtube.com/watch?v=rb3qdb-KvWq8 & see Podcast Knowledge Weekly, Episode "Politics vs. activism: What really changes the world?", September 2021, Link: https://open.spotify.com/episode/7vnBpBbbocXQVYd5qYC03A?si=wQoB6cl-SfGg8K-7cXKzkzQ, accessed 27.04.2023

332   "Why Civil Resistance Works", Erica Chenoweth and Maria

J. Stephan, 2012, Link: https://cup.columbia.edu/book/why-civil-resis-tance-works/9780231156820 & "Why Civil Resistance Works. The Strategic Logic of Nonviolent Conflict", Erica Chenoweth and Maria J. Stephan, International Security, Vol. 33, Issue 1, 2008, accessed 27.04.2023

333   https://twitter.com/BarackObama/status/514461859542351872 & https://unfccc.int/files/meetings/paris_nov_2015/application/pdf/cop21c-mp11_leaders_event_usa.pdf, accessed 27.04.2023

334 https://www.theguardian.com/environment/2024/sep/27/just-stop-oil-activist-phoebe-plummer-jailed-throwing-soup-van-gogh-sunflowers, accessed 28.01.2025

335   DIE ZEIT, for example, counted 4,242 media reports in 2022 in the month after the soup was thrown at a van Gogh painting compared to the 300 in the previous month, which was only char-acterized by street blockades. See also https://www.carbonbrief.org/analysis-how-uk-newspapers-commented-on-energy-and-climate-change-in-2022/?utm_source=cbnewsletter&utm_medium=email&utm_term=2023-02-01&utm_campaign=This+week+UK+newspaper+views+an-alyzed+Met+Office+on+UK+s+record-breaking+2022+Tropical+for-ests+Eruption+warming, accessed 27.04.2023

336   https://klimakommunikation.klimafakten.de/showtime/kapitel-9-bleibe-positiv-sowohl-im-ton-wie-im-inhalt/, accessed 27.04.2023

337   "Radical Flank Effects", H. H. Haines, 2022, Link: https://on-linelibrary.wiley.com/doi/full/10.1002/9780470674871.wbespm174.pub2 & "Radical flanks of social movements can increase support for moderate factions", B. Simpson, et al., 2022, Link: https://academic.oup.com/pnas-nexus/article/1/3/pgac110/6633666?login=true, accessed 27.04.2023. Alleged-ly, a positive such effect was brought about by the protests of the group "Just Stop Oil" in the UK. Quelle: https://www.socialchangelab.org/_files/ugd/503ba4_a184ae5bbce24c228d07eda25566dc13.pdf, Zugriff 27.04.2023

338   "School Strike. Geschichte einer Aktionsform und die Debatte über zivilen Ungehorsam", Simon Teune, 2020, pp. 131-46 in Fridays for Future - Die Jugend gegen den Klimawandel Konturen der weltweiten Protestbewegung, edited by S. Haunss and M. Sommer, Link: https://www.transcript-verlag.de/978-3-8376-5347-2/fridays-for-future-die-jugend-ge-gen-den-klimawandel/, accessed 27.04.2023

339   Conducted by Gallup Organization May 28 - June 2, 1961, and based on personal interviews with a national adult sample of 1,502. Data provided by the Roper Center for Public Opinion Research, from an article by The Washington Post, example links: https://www.crmvet.org/docs/60s_crm_public-opinion.pdf & https://www.norc.org/PDFs/publica-tions/NORCRpt_119.pdf, accessed 27.04.2023

340   https://www.theguardian.com/environment/2024/sep/27/just-

stop-oil-activist-phoebe-plummer-jailed-throwing-soup-van-gogh-sun-flowers, accessed 28.01.2025

341   https://www.theguardian.com/environment/2024/dec/11/britain-leads-the-world-in-cracking-down-on-climate-activism-study-finds, accessed 28.01.2025

342 https://www.theguardian.com/world/2024/dec/20/record-number-of-protesters-will-be-in-uk-prisons-this-christmas & https://www.theguardian.com/environment/2025/jan/29/sixteen-jailed-uk-climate-activists-to-appeal-against-sentences, accessed 28.01.2025

343   https://www.theguardian.com/environment/2023/oct/12/how-criminalisation-is-being-used-to-silence-climate-activists-across-the-world, accessed 28.01.2025

344   "Die Zukunft in unserer Hand: Wie wir die Klimakrise überleben", Christiana Figueres and Tom Rivett-Carnac, C.H.Beck, 2021, ISBN: 978-3-406-77560-4, Link: https://www.chbeck.de/figueres-rivett-carnac-zukunft-hand/product/34266904 & https://www.forbes.com/sites/jeffmcmahon/2020/02/24/former-un-climate-chief-calls-for-civil-disobedience/?sh=78a1ae953214, accessed 27.04.2023

345   "Scientists must act on our own warnings to humanity " , Gardner, et al, 2019, Nature ecology and evolution, Link: https://www.nature.com/articles/s41559-019-0979-y & "Civil disobedience by scientists helps press for urgent climate action" , Capstick, et al, 2022, Nature climate change, Link: https://www.nature.com/articles/s41558-022-01461-y & "From Publications to Public Actions: The Role of Universities in Facilitating Academic Advocacy and Activism in the Climate and Ecological Emergency", Gardner, et al., 2022, Link: https://www.frontiersin.org/articles/10.3389/frsus.2021.679019/full, accessed 27.04.2023

346   https://earth.org/climate-scientists-mobilised-across-the-world-in-largest-scientist-led-civil-disobedience/, accessed 27.04.2023

347   There is not only indignation, but also many who can imagine taking part in civil disobedience actions themselves to achieve more climate protection - for example in the USA: https://climatecommunication.yale.edu/publications/who-is-willing-to-participate-in-non-violent-civil-disobedience-for-the-climate/, accessed 27.04.2023

348   "Protest Movements Could Be More Effective Than the Best Charities", J. Ozden, Stanford Social Innovation Review, 2022, Link: https://ssir.org/articles/entry/protest_movements_could_be_more_effective_than_the_best_charities & "How social movements contribute to staying within the global carbon budget: Evidence from a qualitative meta-analysis of case studies", Thiri, et al., 2022, Ecological Economics, Link: https://www.sciencedirect.com/science/article/pii/S0921800922000180, accessed 01.05.2023

349   https://www.klimafakten.de/meldung/schaden-die-aktionen-der-letzten-generation-dem-klimaschutz-oder-helfen-sie-was-sagt-die, accessed 27.04.2023

350   "The New Climate War", Michael E. Mann, 2021, Link: https://michaelmann.net/books/climate-war & See for comparison with the Second World War: https://www.carbonbrief.org/guest-post-can-green-hydrogen-grow-fast-enough-for-1-5c/?utm_source=cbnewsletter&utm_medium=email&utm_term=2023-02-01&utm_campaign=This+week+Hydrogen+growth+net-zero+and+health+olive+oil+warning+forest+defenders, accessed 14.2.23

351   "Social tipping dynamics for stabilizing Earth's climate by 2050", Illona M. Otto, et al, PNAS, 2020, Link: https://www.pnas.org/doi/full/10.1073/pnas.1900577117 & https://www.carbonbrief.org/tipping-points-how-could-they-shape-the-worlds-response-to-climate-change/?utm_campaign=Carbon%20Brief%20Weekly%20Briefing&utm_content=20220916&utm_medium=email&utm_source=Revue%20Weekly & "Experimental evidence for tipping points in social convention", D. Centola, et al., 2018, Link: https://www.science.org/doi/10.1126/science.aas8827 & "Minority influence in climate change mitigation", J. W. Bolderdijk, L. Jans, 2021, Link: https://www.sciencedirect.com/science/article/pii/S2352250X21000154, accessed 27.04.2023

352   "Social tipping dynamics for stabilizing Earth's climate by 2050", Illona M. Otto, et al, PNAS, 2020, Link: https://www.pnas.org/doi/full/10.1073/pnas.1900577117, Zugriff 27.04.2023

353   https://www.systemiq.earth/breakthrough-effect/ & https://www.theguardian.com/environment/2023/jan/20/super-tipping-points-climate-electric-cars-meat-emissions, accessed 27.04.2023

354   There are many books and articles that have inspired me to write this list. However, I would particularly like to highlight the book "The future in our hands", Christiana Figueres and Tom Rivett-Carnac, Link: https://www.chbeck.de/figueres-rivett-carnac-klimawandel-ueberleben/product/32405828 & the article "Twelve ideas to change the world", Maike Sippel, Link: https://taz.de/Forscherin-ueber-Transformation/!5904272/ & the article "Contributing Effectively in Times of Crisis", Roger Walsh, Link: https://www.whatisemerging.com/opinions/contributing-effectively-in-times-of-crisis, accessed 27.04.2023

355   "Klimagefühle", Lea Dohm and Mareike Schulze, 2022, Link: https://www.droemer-knaur.de/buch/lea-dohm-mareike-schulze-klimagefuehle-9783426286159 & "Klima im Kopf", Katharina van Bronswijk, 2022. link: https://www.oekom.de/buch/klima-im-kopf-9783962383817, accessed 27.04.2023. In addition, the "Psychologists For Future" offer a lot of useful content and resources and even free initial consultations for activists. You

can also find out more about typical climate emotions and how to deal with them from climate psychologist Janna Hoppmann and her social enterprise "ClimateMind" or from environmental psychologist Lea Grosse and the website klimafakten.de.

356   Incidentally, the majority of IPCC authors believe that global warming will reach at least 3°C by 2100. Source: Survey by the scientific journal Nature, 2021, Link: https://www.nature.com/articles/d41586-021-02990-w, Zugriff 28.04.2023

357   From "Im Grunde gut - Eine neue Geschichte der Menschheit", Rowohlt Taschenbuch, 2021, Link: https://www.rowohlt.de/buch/rutger-bregman-im-grunde-gut-9783499004162, Zugriff 27.04.2023

358   https://www.stockholmresilience.org/research/planetary-boundaries.html & "The social shortfall and ecological overshoot of nations", A. L. Fanning, Nature Sustainability, 2021, Link: https://www.nature.com/articles/s41893-021-00799-z, Zugriff 28.04.2023

359   https://www.kateraworth.com/doughnut/, accessed 27.04.2023

360   The "Gaia hypothesis" was put forward in the 1970s by the microbiologist Lynn Margulis and the biophysicist, chemist and physician James Lovelock. Also see "The Reverence for Life" by Albert Schweitzer, summarized for example at https://schweitzer.org/de/die-ehrfurcht-vor-dem-leben/, accessed 27.04.2023

361   See for example https://www.penguinrandomhouse.de/Buch/Zen-und-die-Kunst-die-Welt-zu-retten/Thich-Nhat-Hanh/Lotos/e601543.rhd, accessed 27.04.2023

362   I couldn't find out where this phrase came from. During my research, I often came across the reference "Indian wisdom". A slightly modified version of the phrase was written by environmental activist Wendell Berry in 1971. Source: https://quoteinvestigator.com/2013/01/22/borrow-earth/, accessed 27.04.2023

363   https://www.un.org/sg/en/content/sg/speeches/2022-11-07/secretary-generals-remarks-high-level-opening-of-cop27, accessed 31.01.2025

Photo: Anna Masiukiewicz

Gabriel Baunach was born in 1993 and began engaging with the climate crisis at the age of 14. He studied at Boston University, RWTH Aachen University, the London School of Economics, and Stanford University, earning a degree in Mechanical Engineering with a focus on energy technology. He took part in the UN Climate Conference COP25 in Madrid as part of the UN Climate Secretariat. Since 2020, Gabriel has been producing climate-focused podcasts and giving talks on the climate crisis and the concept of the climate handprint. In 2025, he joined the German NGO Mehr Demokratie, promoting more democratic participation.

www.ingramcontent.com/pod-product-compliance
Lightning Source LLC
Chambersburg PA
CBHW031430270326
41930CB00007B/638